GW00708359

Coronavirus Pandemic

A Survival Guide to Know All the Secrets About Wuhan Coronavirus. Practical Advice to Protect Your Health and That of Your Family from Covid-19 Outbreak

William J. Rodriguez

Table of contents:

Introduction

Coronaviruses are a large structure of viruses, some of which cause human disease and others that circulate among mammals and birds. Animal coronaviruses seldom spread to humans and spread among humans instead. Zoonotic coronaviruses have evolved in recent years, causing human outbreaks such as COVID-19, middle east respiratory syndrome (MERS), severe acute respiratory syndrome (SARS). Human disease sometimes occurs as a respiratory infection, or often as a gastrointestinal infection. The clinical continuum of disease varies from no symptoms or mild symptoms of breathing to severe, rapidly developing pneumonia, acute respiratory distress syndrome, septic shock, or multi-organ failure caused by death.

Centers for disease control react to an outbreak of respiratory disease caused by a coronavirus that was first identified in China and has now been identified worldwide in over 100 countries, including the USA. The virus was named "SARS-CoV-2," and the disease it is causing was renamed "Coronavirus Virus 2019" (abbreviated "COVID-19").

Coronaviruses are a large family of human-common viruses and several different types of animals, including camels, goats, cats, and bats. Animal coronaviruses can rarely infect individuals and then spread to people such as MERS-CoV, SARS-CoV, and now with this new virus (called SARS-CoV-2).

As with MERS-CoV and SARS-CoV, the SARS-CoV-2 virus is a beta coronavirus. These three viruses all originate in bats. The sequences of U.S. patients are similar to those identified initially by China, suggesting a probable single, recent rise of this disease from an animal reservoir.

Early on, most of the patients at the outbreak epicenter in Wuhan Province of Hubei, China, had some link to a large market for seafood and live animals, suggesting animals spread to humans. A rising number of patients reportedly did not have access to the animal markets earlier, indicating spreading from person to person. The spread of person-to-person was subsequently established outside of Hubei and in countries outside China, including the United States. Now, as some parts of the United States do, some overseas destinations have the apparent spread of the virus, causing COVID-19 in the population. Community spread indicates that certain individuals have been infected, and it is not clear how or when they were exposed.

Chapter 1: Wuhan epidemy

China remains the center of a deadly coronavirus outbreak, with cases and deaths highly concentrated in Hubei Province and Wuhan, it's capital. Cases were identified as health officials try to avoid a pandemic in many countries around the world. Chinese authorities have placed at least 20 cities on lockdown as they try to control the outbreak. Travel restrictions were imposed in towns throughout the province of Hubei, where the outbreak occurred. Wholesale transport infrastructure has been closed across the country, affecting at least 36 million people moving around. Infections have been reported in nearly every province in China, as well as in the municipalities of Beijing, Shanghai, Chongqing, and Tianjin.

The epidemic is focused on the city of Wuhan, the provincial capital of Hubei, and home to over 11 million people. Within a matter of days, plans have begun to build a 1,000-bed hospital to ease pressure on existing Wuhan medical facilities overrun with patients. Empty supermarket shelves and barricades have also been rising up.

Wuhan International Airport serves 104 destinations outside China, including 29. From Wuhan to Heathrow, there are three direct flights a week, where arrivals from the affected city can now enter via a separate zone. Screening has started at the three American airports where Wuhan flights land. Checks are also placed at airports in other East Asian countries, which have confirmed the presence of the virus.

1.1 Wuhan china and coronavirus

Zhang Luhua's trouble began at the end of last week when Wuhan's 57-year-old resident began breathing difficulties. She proceeds to Hubei Provincial People's Hospital, one of the city's best hospitals, where she was told by a doctor that she might have a new virus that started causing global concern, but he couldn't personally check her. She was turned away from four other hospitals in the city over the course of the next week until she was told by a doctor at Xinhua Hospital that she definitely had the virus, and would go home and "self-quarantine." "By that time, we had stopped trying as hospitals were either closed or filled with patients with various symptom rates," says her friend, Xie Jing. "We've spent several thousand yuan [renminbi] seeing doctors, but we're still not sure if my mother is a virus carrier," she adds: "Wuhan has done a poor job in disease control." Wuhan, an 11 m area, enjoyed a few moments at the Center for Modern Chinese History. One of central

China's economic centers, it was the capital twice briefly under the nationalists and at the height of the cultural revolution, Mao Zedong took his famous swim in the Yangtze in Wuhan.

Now, Wuhan is at the core of an outbreak of a previously unknown coronavirus: nCoV, as it is called, showed some disturbing parallels with SARS this week, which between 2002 and 2004 killed nearly 800 people. SARS SA has measured global economic costs at $50bn. By mid-afternoon Friday, 26 people had died, and, according to Chinese state television, there were 881 reported cases of nCoV infection, with another 1,073 suspected cases in 20 provinces across the country. Small numbers of cases in the US, Japan, South Korea, Thailand, Singapore, Hong Kong, and Macau have also been confirmed, but no deaths have been reported outside China. Wuhan has been under a near-lockdown with rail, and air connections suspended since Wednesday as the authorities are seeking to curb the spread of infection. Paramilitary police are standing outside the main train station to prevent anyone from entering through— unlike overcrowded hospitals— the streets are relatively quiet ahead of the annual spring festival, China's most important holiday when families come together to celebrate.

Yet some health experts are afraid that travel restrictions on Wuhan and other Hubei cities have appeared too late, given that many of the 100 m Chinese who cross the country to visit relatives each year have already made the trip. For President Xi Jinping's authoritarian government, which is under pressure from the protests in Hong Kong and an opponent's victory in the Taiwanese elections, the sudden happening of a new virus is a very public test of its capability to manage an international crisis— and one that cannot be easily blamed on malignant outside influences. "The myth the Communist Party is trying to promote is that the Chinese government system is fundamentally superior due to its ability to sustain stability and development," says Adam Ni, a former Australian government advisor and editor of the Chinese Neocon newsletter. "Internal shocks will smash the concept of party omnipotence." The outbreak came to the consideration of the authorities on 29 December when four employees were admitted to the hospital with pneumonia symptoms at Huanan Seafood Wholesale Market. "It was detected because China had set up a system specifically designed to collect severe lower respiratory infections," says Tedros Adhanom Ghebreyesus, the World Health Organization's director-general. Within two weeks, Chinese researchers had analyzed the pathogen as a new coronavirus strain and published their full genetic code to colleagues around the world. Both Sars and Mers, a similar coronavirus, originally passed on to humans through interaction with animals— civet cats and camels, respectively— and experts agree that nCoV arose from live animals sold on the market. The source itself is under investigation. While there was mention of snakes as a possibility, many scientists suspect that the virus originated in bats.

Beijing for what Mr. Tedros calls "its engagement and openness" — as opposed to the three months that the country took in 2003 to alert the rest of the world about Sars. Health experts agree that this time their counterparts in Beijing are trying to be as transparent as possible, while local Wuhan authorities may have tried to cover up the outbreak in its early stages. In Wuhan and its neighboring cities, where tens of millions of citizens have been put in partial quarantine, the Chinese system's capacity to marshal resources and issue orders was apparent. Major tourist sites across the country, from Beijing's Forbidden City to Shanghai Disneyland, have shut their doors, while industries that produce surgical masks have continued to run through the holidays.

One 29-year-old resident of Shanghai who declined to be identified says he went to meet his parents in Wuhan on Tuesday and came back home at 10 am on Thursday, just before the travel ban came into effect. He got a knock at the door of his flat in Pudong district at 3 pm. "I was just in my pajamas to find that there was one police officer in my living room, and two medical personnel in full gear," he says. "They tracked me down and approached out to my landlord in order to get the key." He was ordered to stay home for the next week, reporting his body temperature twice a day. He got comments from "haters" who accused him of spreading the virus by leaving Wuhan when he posted a story about his visit to WeChat. "But they have no idea how poorly the local authorities reacted to the incident as a whole," he says. But there could be downsides to Mr. Xi's crisis management leadership style, with centralized decisions leading to inaction among lower officials worried about taking the initiative on sensitive matters. "This is yet another example of his consolidation of power establishing a buck-passing and secrecy-prone political structure," says Jude Blanchette, a China expert at the US Center for Strategic and International Studies. The position of control of the president will not be affected, but the current epidemic, he notes, "carries an exceedingly difficult year." "Xi is looking to create an empire, but then will spend his time battling domestic fires." Unhappiness over the Communist Party's management of the outbreak has so far been aimed primarily at Wuhan's mayor, Zhou Xianwang, with some online critics calling for his resignation for earlier downplaying of the seriousness of the virus. In particular, residents were furious that two activities held in the city in early January— a public dinner of over 40,000 participants and the distribution of 200,000 free tickets to tourist sites — had not been canceled and could have escalated the spread of the virus.

Not all are so critical of the official response. Wang Hongtao, 30, a Wuhan video maker, has spent the last few days studying one of the city's main hospitals. "There was a long queue of patients, but it seemed like everything was kept in order," he says. "I don't think there are hospitals that turn away infected patients." So far, nCoV has killed 3-4 percent of infected people, mainly those with underlying chronic disease, and caused serious illness in about a quarter. This contrasts with a fatality rate of 10 percent in Sars. Although the new virus doesn't easily transfer between people, the Chinese authorities detect restricted human transmission to humans. Jeremy Rossman, a horologist at the University of Kent in the UK, says: "The virus is likely to continue to spread with significant rises in the number of cases in the coming weeks. However, it remains ambiguous if the disease will sustain its current low mortality rate and low incidence of person-to-person transmission, or if the virus will adapt. "While no vaccine is available to prevent infection with nCoV, a crash program is underway to create one. The Oslo-based Coalition for Epidemic Preparedness Innovations is a $750 m alliance between governments, charities, and industry formed in 2017 to avoid possible pandemics. Cepi has initiated three initiatives using different methods to train applicants for the nCoV vaccine, and plans to have one ready for testing on human volunteers within 16 weeks, says Richard Hatchett, the chief executive of the collaboration. "Success is not assured," he says.

A stronger diagnostic test is another method that will help combat the outbreak. Health authorities in China and elsewhere use tests that identify nCoV's genetic sequences, but these tests are slow to obtain and must be conducted in specialist laboratories. One US service, Co-Diagnostics, says preliminary design work for a quicker molecular screening test has been completed. No drug has been developed to treat infections with the coronavirus, although some antivirals can help patients battle against nCoV. While medical science has made outstanding progress in the 17 years after SARS, using the same methods to break the chain of transmission cannot stop the latest coronavirus, says Peter Piot, chief of the London School of Hygiene and Tropical Medicine. "You are isolating patients and ensuring the staff who care for them have full protection from infection," he says. "Then, you track all those that have been in touch with the patient during the incubation period and check for any signs in them." There have been no cases of SARS identified since 2004. If its sister virus can also be cleared out within a year, the planet will be relieved.

Chapter 2: What is coronavirus?

Coronaviruses are a wide family of viruses known to cause diseases ranging from the common cold to more deliberate illnesses such as Middle Eastern Respiratory Syndrome (MERS) and Extreme Acute Respiratory Syndrome (SARS).

These are positive filament RNA viruses which appear under an electron microscope in a corona-like shape. The Coronaviridae subfamily Orthocoronavirinae is categorized into four genera of coronavirus (CoV): Alpha-, Beta-, Delta— and Gamma coronavirus. The betacoronavirus genus is further classified into five subgenera (including the Sarbecovirus).

In the mid-1960s, coronaviruses were identified and are believed to infect humans and certain animals (including birds and mammals). The principal target cells are the respiratory and gastrointestinal epithelial cells.

Seven coronaviruses have been shown to be capable of infecting humans to date: common human coronaviruses are HCoV-OC43 and HCoV-HKU1 (Betacoronavirus) and HCoV-229E and HCoV-NL63 (Alphacoronavirus); mild colds, but also serious infections of the lower respiratory tract.

Many human coronaviruses: SARS-CoV, MERS-CoV, and 2019-nCoV (now named SARS-CoV-2).

2.1 Where did the virus come from?

Chinese public health officials told the World Health Organization at the end of December that they had a problem: an unknown, new virus in the city of Wuhan caused pneumonia-like illness. They soon decided it was a coronavirus, and it was spreading rapidly through and outside Wuhan.

Coronaviruses are widespread in all kinds of animals, and can often develop into forms that can infect humans. Two other coronaviruses have transferred to humans since the turn of the century, triggering the 2002 SARS outbreak, and 2012 MERS outbreak.

Scientists claim that in early December, this new virus first became capable of jumping to humans. Originally it seemed as if the virus first infected people on a Wuhan seafood market and spread out from there. Yet one study of the early cases of the disease released January 24th, found that there was no business interaction with the first patient to become ill. Now, experts seek to track the outbreak back to its source.

The type of animal from which the virus originated is not clear, although one study showed that the new virus's genetic sequence is 96 percent similar to one coronavirus showed in bats. SARS and MERS both emerged in bats.

2.2 So, is this the same as SARS?

No. The new coronavirus (now renamed SARS-CoV-2 and already named 2019-nCoV) belongs to the same family of SARS viruses, but it is not the same virus.

The new Coronavirus, now called COVID-19, responsible for the respiratory disease, is closely related to SARS-CoV and is genetically classified under the Sarbecovirus subgenus Beta coronavirus.

2.3 China lied to the WHO about SARS. Is it lying about this, too?

During the outbreak of SARS, Chinese officials tried to conceal cases from WHO inspectors and restrict details, both internally and externally. This time, officials immediately announced the outbreak of the new virus to the WHO, which in a press conference praised their prompt response and accountability. China is also authorizing a team of WHO experts to assist the ongoing work of Chinese public health authorities, the organization reported on January 28.

The US administration of Health and Human Services also reported that China was more open than it was with SARS. "The level of cooperation between the Chinese government and what we witnessed in 2003 is completely different," Department Secretary Alex Azar said during a press conference.

But critics and Chinese people were cautious: early doubts were raised that Chinese officials were underreporting the number of diseases and classifying as pneumonia deaths that could have been caused by the virus. At the start of the outbreak, Wuhan police also prosecuted residents for spreading what it called online rumors.

(It is necessary to remember that China is not the only country known to conceal the severity of public health problems; in the United States, for example, hundreds of cities have hidden the amount of lead in their common use water supplies.)

2.4 How dangerous is this new virus?

This needs knowledge about how serious a disease is, and how quickly it can spread to assess how "evil" it can be. Epidemiologists also use this technique to determine new types of flu, such as: if a disease is not very serious (and kills just a small number of people), but it is highly transmissible, it can still cause catastrophic consequences— if it affects millions, the small percentage of deaths it causes will still be high.

The WHO called the disease caused by the COVID-19 coronavirus— "co" and "vi" for coronavirus, "d" stood for disease and "19" for the year in which the disease appeared.

The COVID-19 symptoms ranged from moderate to extreme, like those of a cold. Of confirmed cases, about 80 percent are moderate. This is 80 percent of the situations we are aware of. This is also probable that there are several more moderate cases of the disease which have not been reported, which will reduce the number of severe cases. Approximately 5% of cases are critical, and about half of people with serious illness cases tend to die from it.

So far, the risk of fatality for the new disease is around 2 to 3 percent, but it is very soon to tell for sure, and this could change as the epidemic progresses. The SARS fatality rate was between 14 and 15 percent. Many of the deaths in this epidemic were in the elderly and those with underlying health problems, such as hypertension, heart disease, and diabetes. The fatality risk for the latest coronavirus is even higher in that group: for example, it's about 14 percent for people above the age of 80.

2.5 How fast is the virus spreading?

The virus is rapidly spreading across the planet. Since the beginning of January, sick people in China have been infecting others through transmission from person to person. In the confined atmosphere, the latest coronavirus spread rapidly on the Diamond Princess ' cruise ship. Clusters with high numbers with cases have occurred in Italy, Iran, and South Korea, and it is likely that several more cases have not been identified outside of China. Experts say that a wider spread of the virus may not be containable.

Early evidence suggests that the virus, like other coronaviruses, jumps between people in very close contact with each other, and is likely to spread when an infected person is sneezing or coughing.

It's also uncertain when and for how long people become infectious with COVID-19. One analysis of nine people in Germany with moderate cases of the disease showed that early in the course of the disease, they had high levels of the virus in their lungs until they could feel very sick. That could mean people could spread the virus before they realize they've got it.

Chinese officials said they'd seen cases where people with the virus have infected others until they begin displaying symptoms, but there's no clear data to say whether or how much that's going on. Research from China has shown that people without symptoms still have high levels of the virus in their throats and noses, meaning that if they cough or sneeze, they may pass it along. A family in Anyang, China, also appeared to be sickened by an asymptomatic member of the family, a study in JAMA reported.

When this occurs frequently, it will be more difficult to stop the spread of the virus. And even if it's happening, evidence suggests that it probably doesn't affect the epidemic significantly, Maria Van Kerkhove, manager at the WHO's Emerging and Re-Emerging Diseases Unit, said in a press conference. "Asymptomatic transmission is not a major transmission driver," she said.

In a press conference Anthony Fauci, director of the National Institute for Allergy and Infectious Diseases, said the same. "Even though there is some asymptomatic transmission, asymptomatic transmission has never been the cause of outbreaks in all the history of transmitted respiratory disease," he said. "Asymptomatic carriers will not cause an outbreak." The WHO says researchers conclude that each infected person will tend to infect between 1.4 and 2.5 additional people on average, but that is only a tentative estimate. Many study teams have released their own findings, most of which suggest a sick person can infect an average of about two or three individuals.

Those numbers are known as the R0 virus (pronounced "R-naught"). The R0 is the statistical representation of just how far an infection might spread. The greater the number, the more likely it will spread. The SARS R0 was between two and five, for instance. But this doesn't mean that any sick person can potentially infect so many people; quarantines and other measures taken to contain a virus outbreak will cut down the number of people affected by a sick person.

2.6 Can we treat this virus?

There are no approved therapies for COVID-19, but hundreds of trials are continuing to try to find some. One leading candidate is remdesivir, an antiviral medication initially designed for the treatment of Ebola. This is being studied in clinical trials in patients in China and in the US.

A vaccine that can protect people from infection is also being developed by research teams and pharmaceutical companies. It does take a long time to produce vaccines, however. Even if all goes well, it will be about a year to 18 months before one is available, said Anthony Fauci, the National Institute for Allergy and Infectious Diseases director.

2.7 How is China trying to stop the virus?

At the start of the outbreak, China took immediate steps to shut down transportation in Wuhan—home to over 11 million people—and to cancel flights and trains in and out of the area. Just to stop the spread of the virus to other nations, hundreds of other cities were put under successful quarantine, and many canceled celebrations for the Lunar New Year, an enormous holiday in China.

China also stopped public meetings, segregated sick people, monitored their communications actively, and had a dedicated network of hospitals to check for the virus.

The number of new infections identified in China has decreased, signaling to officials at the WHO that transmission is slowing down— and that their containment measures have been working.

Chapter 3: Transmission of coronavirus

You still have concerns about the 2019 coronavirus, as do several others. And one of those issues could be linked to how the virus will spread?

Firstly, a few brief descriptions regarding the coronavirus itself: SARS-CoV-2 is actually the clinical name for this novel coronavirus. It stands for coronavirus 2 of the extreme acute respiratory syndrome.

This emerged from a family of other viruses, which cause respiratory diseases such as severe acute respiratory syndrome (SARS) and respiratory syndrome of the Middle East (MERS).

Since novel coronavirus is a new strain, our immune systems are unfamiliar with this. And for that, there is no vaccine yet.

If a person contracts the virus, the result is COVID-19 disease. It is spread via the respiratory droplets, being a respiratory virus.

Let's take a closer look at how this novel coronavirus spreads from person to person.

3.1 How does it spread from person to person?

According to the Centers for virus Control and Prevention (CDC)Trusted Source, person-to-person contact is thought to be the primary method of transmission for the SARS-CoV-2 virus.

Consider sitting on the bus or in a conference room next to someone with a SARS-CoV-2 infection. Suddenly the person is sneezing or coughing.

If their mouth and nose are not protected, they could theoretically spray you from their nose or mouth with respiratory droplets. The droplets that fall on you will possibly have the virus in them.

Or maybe you meet someone who contracted the virus, and they put their hand on their mouth or nose. When the person is shaking your head, some of the virus is passed into your side.

If you then touch your mouth or nose without first washing your hands, you might accidentally give that virus a point of entry into your own body.

One recent small study indicated the virus could also be found in feces and may contaminate areas such as toilet bowls and sinks in the bathroom. But the researchers acknowledged the likelihood that this may be a mode of transmission that needs further study.

3.2 Can someone spread the virus even if they don't have symptoms?

The World Health Organization (WHO) is recommending right now that the risk of transmitting the novel coronavirus from someone who has no symptoms is very low.

But here's some sobering news: Experts think it's likely that anyone with a novel coronavirus infection might pass it on to others even if they don't display any symptoms, or have such mild symptoms that they don't even realize they're sick.

An individual who contracted the virus is most infectious when they show symptoms— and that's when they're most likely to spread the virus— according to the Center for disease control.

But just before they start displaying signs of the disease itself, someone might be able to pass on the virus. The symptoms may take 2 to 14 days to show up anywhere following exposure to the virus.

One recent study of 181 COVID-19 patients found a median 5-day incubation period, with more than 97 percent displaying symptoms 11.5 days after exposure to the virus.

3.3 Can you pick it up from infected surfaces?

Think about all the regularly touched surfaces where germs can chuckle: kitchen counters, bathroom counters, door buttons, elevator buttons, refrigerator handle, staircase handrails. The list continues and goes on.

Experts are not sure how long the novel coronavirus will live on these surfaces. But if the virus acts like other, similar viruses, the time for survival may vary from several hours to several days.

The surface texture, room temperature, and ambient humidity can play a role in how long the virus will live on a surface.

However, because we don't know for sure if you think a surface could be dirty, thoroughly clean it with a disinfectant. To this end, a diluted bleach solution or a disinfectant approved by EPA is possibly the most effective cleaner.

And if anyone is sick in your house, clean those surfaces regularly. Afterward, try to wash your hands thoroughly.

3.4 This is how the disease progresses: (Day 7 is the worst.)

"We now have a better understanding of the overall time course for the illness from published studies," Chiu said. When a person is exposed and infected, the time of incubation until the onset of symptoms is approximately five days, although this can range from two to 11 days. At first, flu-like symptoms are always mild, and some patients recover without the symptoms becoming more severe. But for a subset that gets worse, day four after the onset of symptoms is usually when they seek medical treatment as they experience shortness of breath and early pneumonia, Chiu said, and by day seven, they may become seriously ill. Following day 11, most remaining patients are on their way to recovery.

3.5 Even those who recover from COVID-19 might not be immune forever.

"Unfortunately, we still don't know if the body's immune response would protect you from subsequent infection," Chiu said. Exposure to the four seasonal human coronaviruses (which causes the common cold) is known to produce immunity to those specific viruses. In such cases, the immunity lasts longer than seasonal influenza, but is not definitely permanent, Chiu said.

3.6 A single negative test may not rule out infection.

The screening test currently available is a PCR test developed by the Centers for disease control, which searches for RNA from the virus. Nevertheless, hospitalized patients infected with the latest coronavirus will have test results that differ from day to day as the amount of virus the body produces will change over the course of the disease.

Repeat testing can be appropriate to decide if a defendant was contaminated or whether a patient is no longer infectious. "The message that you take home is that a test that looks at a single point of time is not enough to rule out infection.

Data from the Washington State case also shows that the extent of the disease does not always equate with rates of the body's virus—suggesting someone may be very contagious without appearing to be very sick. "This is why there is fear that the outbreak might be exacerbated by patients who are minimally symptomatic simply because they don't feel sick enough to go to the hospital.

Chiu's laboratory is designing, in partnership with the startup Mammoth Biosciences, a quick diagnostic test that could track the disease faster and more broadly. The latest test is a test strip that changes color and uses CRISPR to detect viral RNA, which can be performed in 30 minutes to an hour. "We were able to run this rapid test on both control samples and patient samples, and it seems to work," Chiu said. He aims to refine the test in such a way that it can be performed by anyone and implemented in areas with low resources.

3.7 Health care workers are taking maximal precautions when treating COVID-19 patients.

Health care staff wear personal protective equipment such as gowns, boots, face covers, and N95 respirator masks to protect against contamination by the latest coronavirus while they are in the same room as a patient who is in isolation. "All health care workers get standard N95 mask fittings to ensure they get worn properly," Chiu said. These protections are intended to protect against touch, droplet, and airborne transmission. Additional airborne precautions are taken in the health care environment as some surgical procedures, such as endotracheal intubation, can aerosolize the secretions. Chiu said that objects brought into the room are ideally disposable by the Center for disease control instructions, or if not, are disinfected until they are removed from the room, and the whole room is disinfected after the patient is discharged from hospital.

Chapter 4: Symptoms and myths of coronavirus

According to Harvard T.H. Chan academy of Public Health epidemiologist, if effective controls are not put in place, COVID-19
could ultimately infect between 40 percent and 70 percent of the population worldwide in the coming year.

4.1 What are the symptoms?

Each day, the doctors know new things about this infection. So far, we know that COVID-19 can not cause any symptoms at first.

Once you experience symptoms, you will bear the virus for two days, or up to 2 weeks.

Some common symptoms specifically associated with COVID-19 include:

- shortness of breath
- cough that becomes more severe over time
- low-grade fever that increases gradually in temperature

4.2 COVID-19 versus the flu

Coronavirus of 2019 is much more dangerous than the seasonal flu.

An approximate 0.06 to 0.1 percent of people who contracted flu in the United States during the 2019–2020 flu season died (as of February 2020), compared with about 3 percent of those with a reported COVID-19 case in the United States.

Here are a few common flu symptoms:

- cough
- runny or stuffy nose
- sneezing
- sore throat
- fever
- headache
- tiredness
- chills
- body aches

4.3 Coronavirus may be most infectious when symptoms are mildest

According to a small study, people infected with the novel coronavirus shed large quantities of the virus early in their illness and are likely to become less infectious as the disease wears.

The research, posted to the database medRxiv on Sunday (March 8), is still preliminary, as it has not yet been peer-reviewed and as it only included nine participants. Even, it might explain that the new virus spreads so easily: while experiencing only mild, cold-like symptoms, many people might be at their most contagious.

"This is in sharp contrast to SARS," a similar disease caused by another coronavirus, noted the scientists. Viral shedding in SARS patients peaked around seven to 10 days into the disease, as the infection escalates from the upper respiratory tract into the deep lung tissue. For seven COVID-19 patients, the disease caused by the new virus, "peak concentrations were reached before Day 5 and more than 1,000 times higher" than those seen in patients with SARS, the authors wrote.

This peak later occurred in two patients whose lung infections had advanced, causing the first symptoms of pneumonia. For these extreme cases, viral shedding occurred at around Day 10 or 11. In the mild cases, after Day 5, the viral shedding steadily dipped, and by Day 10, patients were likely no longer contagious, the authors noted.

"Based on the results reported, early discharge with subsequent home isolation should be chosen for patients with symptoms beyond Day 10," given that swab samples from their throats contain fewer than 100,000 copies of viral genetic material per milliliter, the authors wrote.

"It is a very significant contribution to understanding both the natural history of COVID-19 infectious disease and the effects of viral shedding on public health," Michael Oster Holm, chief of the department for Infectious virus Research and Policy at the University of Minnesota, told Stat News.

The researchers carried out their study by taking swabs from the patients' noses and throats and also analyzing their blood, urine, vomit, and sputum— a mixture of saliva and mucus that builds up during infection in the respiratory tract. To determine how much of the virus was present at different stages of the disease, the team examined each sample for bits of viral genetic material called RNA.

Researchers were monitoring the virus' rise and fall over time. Nevertheless, viral load cannot indicate whether patients remained infectious because RNA from the virus can be present but not usable in human tissue. The researchers isolated samples of the virus during the study to find out who was infectious and when, and attempted to develop them in the lab.

Researchers found they were able to develop virus from the samples of the mouth, nose, and sputum obtained early in the course of illness, but samples taken from patients with mild cases did not show any viral growth after Day 8. The shift indicates some patients were less contagious. Nevertheless, given their progress, they have always screened for the virus as "positive." The finding can help explain Chinese reports indicating that after COVID-19 symptoms clear up, the virus will remain in the body for at least two weeks.

In the new study, the team was unable to develop a virus from any blood or urine samples obtained during the analysis, nor was it able to grow virus from stool. The stool analysis was based on 13 samples collected from four patients between Day 6 and Day 12, as these contained the largest quantities of viral RNA and allowed researchers to isolate samples. An earlier study from China and the World Health Organization indicated that "viable virus" may be retrieved from the stool of infected people, but it was unknown if such fragments led to the spread of disease.

The authors noted that, because the new study is focused on a select number of relatively mild cases, further work is required to determine how stool might contribute to COVID-19 transmission.

Notably, the team found antibodies in each of the patients between Day 6 and Day 12, indicating that soon after exposure, the immune system begins developing a response against the pathogen. Scientists still don't know whether this rapid immune response occurs in most patients, particularly those with more serious infections.

How is pneumonia related to COVID-19?

Patients experience pneumonia in more serious cases of COVID-19, meaning their lungs begin to fill with pockets of pus or fluid. This results in extreme shortness of breath as well as severe coughing.

According to Paul Biddinger, director of the emergency preparedness research, assessment, and practice program at the Harvard T.H. Chan Academy of Public Health, who spoke on a university webcast on March 2nd, testing for the virus that causes COVID-19 in the United States is generally limited to people with serious symptoms. That means testing at the first sign of a fever or sniffle is not acceptable. Biddinger added that the search for medical treatment for a mild disease could also potentially spread the disease or contribute to new diseases being caught in the hospital or clinic.

If you become sick with these signs of illness and reside in or have traveled to a spreading area of COVID-19, which now includes areas of the U.S., the Centers for disease control suggests contacting your doctor first, after that visit a clinic. Doctors working with state health departments and the Centers for Disease Control have to decide who the new virus should be screened for. The CDC also suggests, however, that people with COVI-19 or any respiratory disease carefully monitor their symptoms. Worsening breath shortages is a reason to seek medical care, particularly for the elderly or those with underlying health conditions.

4.4 Coronavirus myths busted by science

Myth: Face masks will defend you against viruses
Normal surgical masks cannot protect you against SARS-CoV-2, as they are not built to filter out viral pieces and do not lay flush to the face, previously stated by Live Science. This being said, surgical masks will help prevent infected persons from further transmitting the virus by covering any respiratory droplets that might be expelled from their mouths.

Unique respirators called "N95 respirators" have been shown to dramatically reduce the spread of the virus among medical staff within health care facilities. People need training to properly fit N95 respirators around their noses, cheeks, and chins to ensure that no air can go out around the edges of the mask, and after each use, wearers do need to learn to test the equipment for harm.

Myth: You're way less likely to get this than the flu

Needless to say. Scientists measure their "simple number of reproductions," or R0 (pronounced R-nought), to determine how quickly a virus spread. R0 estimates the number of individuals who can catch a given bug from a single infected human, previously stated by Live Science. Now, the R0 for SARS-CoV-2, the virus that causes the COVID-19 disease, is estimated at about 2.2, meaning that, on average, a single infected person will infect 2.2 others. The flu has a R0 of 1.3, by comparison.

More notably, maybe, although there is no vaccine to prevent COVID-19, the seasonal flu vaccine prevents influenza fairly well, even though its formulation doesn't suit the circulating viral strains perfectly.

Myth: The virus is actually a mutation of the common cold

No, that is not. Coronavirus is a large virus family containing several different diseases. SARS-CoV-2 exhibits similarities with other coronaviruses, of which four may induce common cold. All five viruses have spiky surface projections and use so-called spike proteins to infect host cells. The four cold coronaviruses— named 229E, NL63, OC43, and HKU1 — all use humans as their primary hosts, though. SARS-CoV-2 resembles about 90 percent of its genetic material with bat-infecting coronaviruses, indicating that the origin of this virus was in bats and then jumped to humans.

There is evidence that the virus went through an intermediate species before infecting humans. Similarly, on its way into humans, the SARS virus jumped from bats to civets (small, nocturnal mammals), while MERS affected camels before growing to humans.

Myth: The virus was definitely produced in a laboratory

No data suggests the virus is man-made. SARS-CoV-2 closely resembles two other coronaviruses, SARS-CoV and MERS-CoV, that have triggered outbreaks in recent decades, and all three viruses appear to have originated in bats. In short, SARS-CoV-2's features fall in line with what we learn about other naturally occurring coronaviruses that made the transition from animals to humans.

Myth: Getting COVID-19 is a death sentence

It is not real. According to a report reported by the Chinese Center for Disease Control and Prevention, about 81 percent of people diagnosed with the coronavirus have mild cases of COVID-19. About 13.8 percent experience serious illness, which means they have shortness of breath or require supplemental oxygen, and about 4.7 percent are critical, which means they face respiratory failure, multi-organ failure, or septic shock. So far, the data show that only about 2.3 percent of COVID-19-infected people die from the virus. People who are senior citizens or have existing health problems tend to be at higher risk for serious illness or complications. While caution is not required, people should take measures to prepare and protect themselves and others against the new coronavirus.

Myth: Pets can spread the new coronavirus

Maybe not to humans. According to The South China Post, one dog in China contracted a "low-level infection" from its owner, who has a reported COVID-19 outbreak, suggesting dogs could be vulnerable to having the virus from humans. The infected Pomeranian did not become sick or exhibit signs of disease, and there is no evidence to indicate that the animal could infect humans.

During an outbreak in 2003, animal health specialist Vanessa Barrs, of City University, told the Post, multiple dogs and cats tested positive for a common virus, SARS-CoV. "Previous SARS experience indicates that cats and dogs do not get sick or pass on the virus to humans," she said. "Importantly, there was no clue of viral transmission from pet dogs or cats to humans." The Department for Disease Control and Prevention (CDC) recommends that people with COVID-19 should have someone else walking and taking care of their animals while they are sick. And people just have to wash their hands after cuddling with animals anyway because, according to the CDC, the companion pets can transmit certain diseases to people.

Myth: There won't be lockdowns or school suspensions in the US

There is no guarantee, but closures to schools are a common tool used by public health officials to slow or stop the spread of contagious diseases. For instance, 1,300 schools in the U.S. closed to reduce the spread of the disease during the swine flu pandemic of 2009, according to a 2017 study by the Journal of Health Policy, and Law. According to the report, CDC guidelines at the time recommended that schools close for between 7 and 14 days.

Although the coronavirus is a particular disease with a different time of incubation, transmissibility, and severity of symptoms, it is likely that there will be at least some closures of education. Dr. Amesh Adalja, an infectious disease specialist at the Johns Hopkins Center for Health Protection in Baltimore, previously told Live Science if we know later that children are not the primary vectors for disease, the strategy can change. Either way, the likelihood of school closures should be prepared, and contingency treatment should be formulated if appropriate.

Other options include lockdowns, quarantines, and isolation. Under category 361 of the Public Health Service Act (42 U.S. Code § 264), it is appropriate for the federal government to take these measures to quench disease transmission from either outside the country or within states. Similar authorities may also apply to state and local governments.

Myth: Kids can't catch the coronavirus

Children can definitely catch COVID-19, although initial reports in children suggested fewer cases compared to adults. For example, a Chinese study released in February from the province of Hubei found that of more than 44,000 COVID-19 cases, only around 2.2 percent included children under the age of 19.

Newer studies, however, suggest that children are as likely to become infected as adults. Researchers evaluated data from more than 1,500 people in Shenzhen in a study published on March 5, and found that children potentially exposed to the virus were just as likely to become infected as adults, Nature News says. Regardless of age, roughly 7 to 8 percent of COVID-19 contacts subsequently tested positive for the virus.

Even when children are sick, they are less likely to experience serious illness, previously stated by Live Science.

Myth: If you have coronavirus, "you'll know."

No, you're not going to. COVID-19 is responsible for a wide variety of effects, many of which occur in many respiratory illnesses e.g., common cold and the flu. Specifically, typical COVID-19 symptoms include fever, cough, and breathing problems, and rarer symptoms include dizziness, nausea, vomiting, and a runny nose. In extreme cases, the disease may develop into a serious pneumonia-like disease— but infected people may display no symptoms at all early on.

U.S. health authorities have now urged the American public to brace for an outbreak, meaning those who haven't traveled to the affected countries or had contact with people who have recently traveled may begin to catch the virus. If the epidemic continues in the U.S., departments of state and local health will have updates about where and where the virus has spread. If you live in a region affected with a virus and begin experiencing weakness, high fever, shortness of breath, or lethargy of breath, or have underlying conditions and milder symptoms of the disease, experts told Live Science that you should seek medical assistance at the nearest hospital.

From there, you may be checked up or tested for the virus, although the CDC has not made the diagnostic examination available widely available yet.

Myth: Coronavirus is less lethal than influenza

So far, the coronavirus seems to be more lethal than the flu. But there is still a lot of doubt about the virus's mortality rate. Usually, annual flu has a mortality rate of around 0.1 percent in the U.S. So far, according to the CDC, there is a mortality rate of 0.05 percent among those who caught this year's U.S. flu virus.

Recent data, in contrast, show that COVID-19 has a mortality rate more than 20 times higher, around 2.3 percent, according to a study published by the China CDC Weekly on February 18. According to a previous Live Science study, the death rate differed by different factors such as location and an individual's age.

But those figures are changing constantly and may not reflect the real mortality rate. It's not clear if the case counts are correctly recorded in China, particularly because, according to STAT News, they changed the way they described cases midway through. There could be many temperate or asymptomatic cases which, they wrote, were not included in the total sample size.

Extra vitamin C will save you from getting COVID-19
Researchers have yet to find evidence that supplements with vitamin C can protect people from the COVID-19 infection. In fact, taking extra vitamin C doesn't even ward off the common cold for most people, though if you catch one, it can shorten the duration of a cold.

That being said, vitamin C plays an important human body function and promotes proper immune function. The vitamin, as an antioxidant, neutralizes charged particles called free radicals, which can damage the body's tissues. It also helps the body to synthesize hormones, build collagen and seal off pathogens from vulnerable connective tissue.

So yeah, if you wish to maintain a strong immune system, vitamin C will completely be included in your regular diet. But mega dosing on antioxidants is unlikely to lower the risk of catching COVID-19 and, if you are infected, will give you a "modest" advantage against the virus. No evidence suggests that other so-called immune-enhancing supplements— such as zinc, green tea, or echinacea— either help to prevent COVID-19.

Be careful of products advertised for the new coronavirus as treatments or cures, since the outbreak of COVID-19 started in the USA. The Food and Drug Authority (FDA) and the Federal Trade Commission (FTC) have already issued letters of warning to seven companies for the sale of fraudulent products that promise to cure, treat or prevent the virus.

Myth: It's not safe to receive a package from China

According to the World Health Organization, having letters or packages from China is safe. Past work has shown that coronaviruses don't live on items like letters or packages for long. Researchers think that this new coronavirus is likely to live poorly on surfaces based on what we learn about previous coronaviruses such as MERS-CoV and SARS-CoV.

A previous research showed that, according to a report published Feb. 6 in The Journal of Hospital Infection, these associated coronaviruses might persist on surfaces such as metal, glass, or plastic for as long as nine days. But the surfaces in the package aren't suitable for the survival of the virus.

In order for a virus to stay active, it needs a combination of different environmental conditions such as temperature, lack of UV exposure and humidity— a combination that you won't get in shipping packages, according to Dr. Amesh A. Adalja, senior scholar, Johns Hopkins Center for Health Protection, who spoke with Tom's Hardware sister site of Live Science.

And therefore, "there is generally a very low chance of spreading from goods or containers which are delivered at ambient temperatures over a period of days or weeks," according to the CDC. "There is currently no clue to support the transmission of COVID-19 related to imported goods, and no cases of COVID-19 related to imported goods have been recorded in the United States." However, the coronavirus is believed to be most widely transmitted by respiratory droplets.

Myth: You can get the coronavirus if you eat at Chinese restaurants in the US

No, you just can't. You will also have to avoid Italian, Korean, Japanese, and Iranian restaurants by that argument, provided that those countries were all facing an epidemic as well. The new coronavirus does not just affect people of Chinese descent.

Chapter 5: Complications faced by the global economy due to COVID-19

So far, the illness caused by the coronavirus known as COVID-19 has been identified among less than 0.0008 percent of humans on Earth. But the whole world has been affected, due to the distribution of disease and money.

Chinese manufacturing towns like Wuhan, the epicenter of the outbreak, are closely intertwined with the world's supply chains. This means both the epidemic and the containment measures introduced to regulate it (taking, for example, the quarantine currently in force for 70 million people) would have a drastic impact on companies across diverse industries.

Any business that brings products in from China—including Apple and Walmart—has to worry about slowdowns in production and distribution. That's due in part to supply chains being less linear than they look. Networks of production also have complex interrelationships that go back and forth across borders. An American retailer may contract with just one Chinese corporation, but that firm may operate as a general contractor, in turn, pulling products from several suppliers or contracting out work to a shifting factory list. For example, in

2018, more than 1,000 facilities had been involved in producing Apple products in some way.

Elsewhere, Chinese exporters–such as Brazilian ranchers and Chilean winemakers–face a major decrease in demand from China. Within China, the economic downturn is spreading beyond the manufacturing sectors; as a result of the epidemic, even a media company said it laid off 500 employees.

What makes this all so deviating is that the economic implications of coronavirus are known to a multitude of evidence, but the arrival outside of China of those implications would be delayed, and their significance is unknown. It doesn't help that critics within and outside China have for years doubted the reliability of official country statistics. Yet local news also offers reasons to question the statistics of coronaviruses.

What Target managers think about today will likely be turning up in April for customers. You would assume that stock markets should at least "price in" the problems, but share prices are at record highs. The coronavirus has possibly already dealt with many of its economic blows— and now those shocks will flow through the networks linking China to the rest of the world economy.

Most of the consequences would be material: On supermarket shelves, there will be fewer products, most prices will increase, product creation will slow down. Yet some of the effect will come from the data reflecting the experience of the past two months, most of which has yet to be tabulated, and an important source of latency. Businesses and policymakers need information to understand what's going on in the world. For example, the U.S. government conducts a comprehensive data-gathering operation: The Economic Analysis Bureau, the Bureau of Labor Statistics, several Census surveys services, the National Agricultural Statistics Service, the Economic Research Service, and many more. The data published by those organizations take time to represent on-the-grounded trade. That may not be important under normal conditions. But when the economy experiences a shock that is transforming the globe, the world's statistical windows can be increasingly out of touch with reality.

The data points which can be marshaled to make sense of the macroeconomic image are not good for now. Chinese oil demand fell 20 percent earlier this month, "probably the biggest demand shock on the oil market since the 2008-2009 worldwide financial crisis, and the most sudden since the attacks of Sept. 11," as Bloomberg put it. With some major Chinese towns under different lockdown models, the total number of cars and trucks on the road has declined. Nor are factories operating at full capacity.

According to Morgan Stanley, pollution near Shanghai, a reliable and hard-to-fake predictor of economic growth, has plummeted. Container ships sail at loads lower than average. Prices have collapsed for bulk carriers carrying iron ore and coal. One expert told the Financial Times that the coronavirus "will have a greater impact on the global tech supply chain than SARS and generate more confusion than the U.S.-China trade war." That very trade war caused some companies to move their supply chains to other Asian countries, but China remains the beating heart of production and distribution for the world's products. "Suddenly, all supply chains seem fragile, since too many Chinese supply chains within supply chains based on each other for raw materials and parts," Rosemary Coates, a supply chain consultant, wrote in Logistics Management, a trade newspaper. "The tiny valve you buy for your U.S.-made product inside a motor is manufactured in China. So are the rare earth elements that you need to manufacture magnets and electronics. "The impacts can also differ widely from province to province and even from factory to factory depending on how local governments control their regions, CNBC's chief of the Beijing office, Eunice Yoon, noted.

The slow industrial movement out of China has also left some industries with depleted inventories, like toy making. Companies that spent the last year developing new production networks in other Asian countries are more robust in the long run but may not have enough product to sell at this specific moment.

Less common side effects have also sprung up. When Indonesia's president called for fiscal spending to avoid an economic recession, Indonesian garlic prices soared 70 percent, presumably because Chinese buyers were buying up the folk cure in bulk. Even small ripples may have some effect: In Australia, where after the summer holiday, students from China were unable to return to college, universities pushed back their starting dates, which hurt the businesses around them. The question is, do all these little issues and difficulties add up to something more serious than irritation.

Consider then the economic slowdown's political implications. What if the coronavirus epidemic slows China's economic growth enough to destabilize control of the Communist Party? Bill Bishop, a long-time China analyst, wrote that the outbreak is "the closest thing to an experiential crisis for Xi [Jinping] and the Party that I think we've seen since 1989." The coronavirus is a fascinating test of the complex relationships that hold up the economy today. In our environment, the flow of knowledge is much faster than goods. This means we can experience a big world occurrence, weeks before its effect is quantified in quarantine zone tweets and videos. It is an awkward and unusual place, like recognizing that an earthquake has struck but not recognizing whether a tsunami is on the way. An upshot is likely for Americans, though: Even though the worst epidemic is over — and might not be — bad economic news could be in our future. Or if millions of Chinese workers can be displaced, and the American economy can plow it all through without a glitch, then it may be time to revisit how strong the "Chimerica" link really is.

Chapter 6: Virus Protection: Disinfectants and Face Masks

Coronavirus transmission occurs much more commonly through respiratory droplets than through fomites. Current evidence suggests that novel coronavirus on surfaces made from a variety of materials could remain viable for hours to days. Cleaning visibly dirty surfaces followed by disinfection is a best practice measure in households and community settings for the prevention of COVID-19 and other viral respiratory illnesses.

6.1 How to Keep Your Home Free of Coronavirus Germs?

Staying healthy from the latest coronavirus, for many people, means staying home. However, contagious germs may also be residing in your house.

To reduce the risk of becoming sick, the Centers for Disease Control and Prevention recommends that steps be taken to disinfect high-touch surfaces, such as countertops, doorknobs, mobile phones, and flush handles for toilets, as certain pathogens can live on surfaces for several hours.

But many people don't properly disinfect, says Brian Sansoni, head of communications for the American Cleaning Institute, a Washington trade organization that represents manufacturers of goods. First, before you disinfect, you can need to clean—removing grease or grime. Second, the disinfectant must remain on the surface, often for several minutes, until it dries or is cleaned off. "Check the wait times label to ensure the virus kill is successful," says Mr. Sansoni.

Bleach and other cleaning items have been on short supply in recent days. Mr. Sansoni says factories have been cranking up production to meet demand. That being said, he cautions against overuse of chemical cleaners and, worse, combining cleaners in hopes of improving their effectiveness.

"Panic-cleaning is no requirement," he says. To clean up and sanitize the right way, just read the labels on everyday items. "We should do as they are expected to do." Here are some other recommendations for staying healthy at home: The CDC suggests for at least 20 seconds to wash hands thoroughly with soap and water. Using hand sanitizers, which are at least 60 percent alcohol as a backup.

Recently, the Environmental Protection Agency released a list of disinfectants licensed for killing coronavirus. Look for products such as wipes, sprays, and concentrates which say "disinfectant" on the label and have an EPA registration number for surface cleaning. These are required to meet government safety and efficacy specifications.

The Centers for disease control suggest combining a quarter-cup of household chlorine bleach with one-gallon cool water for a homemade disinfectant.

Rinse them with water before use, after disinfecting food-prep surfaces such as cutting boards and countertops.

Using detergent and bleach (for white loads) for washing, or peroxide, or color-safe bleach (for colors) to remove germs. (Please read clothing labels to avoid damaging garments.) Some washing machines have sanitized or steamed settings that destroy germs to improve the effect. It is also efficient to dry laundry over the hot cycle of the dryer for 45 minutes.

Operate dishwashers on the sanitization cycle where possible. Machines accredited by NSF International, formerly known as the National Sanitation Foundation, have to hit a final rinse temperature of 65.5 Celsius and achieve a bacteria reduction of at least 99.999 percent while working on that cycle.

Household air purifiers and filters which advertise the ability to destroy or catch viruses can be useful but should not be a cleaning substitute. Some purifiers use ultraviolet light, which has been shown to have antiseptic effects, but their overall performance can vary depending on their design, according to an EPA 2018 technical review of residential air cleaners. Although some filters advertise the ability to trap things like viruses, smoke, and common allergens, they don't necessarily kill microorganisms.

6.2 Here's a list of disinfectants you can use against coronavirus

Thursday, the federal government released a five-page list of chemicals and products that it claims are powerful enough to fend off "harder-to-kill" viruses than SARS-CoV-2, the disease-causing virus.

"The use of the right disinfectant is a vital part of preventing and minimizing disease spread along with other crucial aspects including hand washing," said EPA Administrator Andrew Wheeler in a statement.

"There is no greater priority for the Trump Administration than protecting Americans ' health and safety. EPA offers this critical information on disinfectant products in a clear and open way to help minimize the spread of COVID-19," he said.

The Environmental protection agency says it is best to follow the instructions on the packaging of the disinfectant and pay attention to how long the product will be on the surface that you are cleaning.

In a declaration to CNN, the EPA said companies could apply for an "emerging claim of pathogens" based on previously approved claims for more difficult-to-kill viruses. The department is evaluating them and deciding if the organization can make the argument safely.

Nonetheless, one significant thing to note: According to the Centers for Disease Control and Prevention, handwashing with soap and water is also the safest way to avoid virus transmission.

That is because from what we know so far, people and their respiratory droplets are thought to transmit the novel coronavirus mainly— think coughs, sneezes, spit.
To put it another way, transmission from person to person is most common.
Although it is possible that people who touch virus-contaminated surfaces or objects and then touch their mouths or eyes may become infected as well, this may not be the primary way the virus spreads, the CDC said. Thus, wipes with disinfectants can only go so far.

Here are some of the disinfectants listed on the EPA's list:

- Clorox Multi-Surface Cleaner + Bleach
- Clorox Disinfecting Wipes
- Clorox Commercial Solutions® Clorox® Disinfecting Spray
- Lysol brand Heavy-Duty Cleaner Disinfectant Concentrate
- Lysol Disinfectant Max Cover Mist
- Lysol brand Clean & Fresh Multi-Surface Cleaner
- Purell Professional Surface Disinfectant Wipes
- Sani-Prime Germicidal Spray

6.3 General Recommendations for Routine Cleaning and Disinfection of Households:

Community members may perform routine cleaning of frequently touched surfaces (e.g., tables, doorknobs, handles, light switches, desks, toilets, faucets, sinks) with household cleaners and surface-appropriate EPA-registered disinfectants, following label instructions. Labels contain instructions to use the cleaning product safely and effectively, including precautions you should take while applying the product, such as wearing gloves and ensuring that you have good ventilation during product use.

- Family members should be informed about the COVID-19 signs to avoid COVID-19 from spreading in households.
- Clean and disinfect high-surfaces daily in common areas of the household (e.g., tables, hard-chairs, door buttons, light switches, remotes, handles, desks, toilets, sinks) o In the sick person's bedroom/bathroom: consider reducing the frequency of cleaning to what is needed (e.g., soiled items and surfaces) to avoid unnecessary contact with the ill person.
- An ill person should stay in a separate room and away from other people in their homes as much as possible, following guidance from home care.
- The caregiver may provide personal cleaning of the room and bathroom for an ill person unless the room is occupied by a child or by another person for whom such supplies may not be necessary. These supplies include tissues, towels made from paper, cleaners, and disinfectants registered with EPA.
- If there is no separate toilet, the toilet should be washed and disinfected by an ill person after use. If this is not possible, the caretaker should wait for the high-touch surfaces to be cleaned and disinfected as long as it is practical after use by an ill person.

- Family members should follow guidelines about home care when dealing with suspected/confirmed COVID-19 individuals and their isolation rooms/bathrooms.

How to clean and disinfect:

Surfaces:
- Wear plastic gloves as surfaces are washed and disinfected. For every cleaning, the gloves should be discarded. By using reusable gloves, these gloves should be devoted to COVID-19 surface cleaning and disinfection, and should not be used for other purposes. Consult the manufacturer's instructions for the items used for cleaning and disinfection. Clean your hands immediately after removing gloves.
- If the surfaces are dirty, use a detergent or soap and water to clean them before disinfection.
- Diluted household bleach solutions, alcohol solutions with slightly 70 percent alcohol, and most popular household disinfectants registered with EPA should be successful for disinfection.
- Diluted household bleach solutions may be used for surface use where appropriate. Follow manufacturer's design guidelines and adequate ventilation. Verify that the drug is not past its expiry date. Never pair the kitchen

bleach or some other cleanser with ammonia. Once properly diluted, unexpired household bleach can become effective against coronaviruses.

- Prepare a bleach solution by mixing: approximately 5 tablespoons (1/3rd cup) of bleach per gallon of water or approximately 4 teaspoons of bleach per quarter of water. O Products with EPA-approved emerging viral pathogens are expected to be successful against COVID-19 based on data for more difficult viral killing. Follow the guidelines of the manufacturer for all cleaning and disinfection products (e.g., concentration, method of application, contact time, etc.).

- Eliminate visible contaminants when present and clean with suitable cleaners specified for use on such surfaces for soft surfaces, such as carpeted floors, rugs, and drapes. After cleaning: o Launder products according to the manufacturer's instructions, as necessary. Where possible, launder items completely using the warmest acceptable water setting for the items and dry goods, or use products with claims of emerging viral pathogens licensed by the EPA that are ideal for porous surfaces.

Dress, sheets, linens, and other things that go in the laundry:

- Wear plastic gloves when treating a sick person's dirty laundry, and remove them after each use. If reusable gloves are used, such gloves should be reserved for COVID-19 surface cleaning and disinfection and may not be used for other household purposes. Clean your hands immediately after removing gloves.

- If no gloves are used to treat dirty laundry, then wash your hands afterward.
- Don't shake dirty laundry, if possible. This will minimize the chance of virus dispersal through the air.
- Launder products as needed, in compliance with the instructions of the manufacturer. Where possible, completely launder items using the warmest suitable water setting for the items and dry items. Dirty laundry from a diseased person may be washed with things from other people.
- Clean and disinfect surface clothing hampers, as directed above. Consider adding a bag liner, if possible, that is either disposable (can be thrown away) or can be washed.

Hand hygiene and other preventive measures:
- Family members will clean their hands regularly by washing their hands with soap and water for 20 seconds, including immediately after removing gloves and after contact with the ill person. If water and soap are not available and hands are not clearly dirty, a hand sanitizer based on alcohol can be used, which contains at least 60

percent alcohol. If hands are clearly dirty, however, wash hands with soap and water at all times.

- Household members should take usual preventive measures at work and at home, including prescribed grooming of the hand and avoid touching the eyes, nose, or mouth with unwashed hands.

Other key periods for washing hands include:

- After blowing your nose, coughing or sneezing
- After using the bathroom
- Before eating or preparing food
- After contact with animals or pets
- Before and after providing daily treatment for someone else in need of support (e.g., a child)

Other considerations:

- If possible, the ill person will eat/nourish in their house. Items used by the non-disposable food service should be treated with gloves and cleaned with hot water or in a dishwasher. Clean hands after handling products used for the foodservice.
- Dedicate a lined trash can to the sick person whenever possible. Using gloves to cut garbage containers, treat waste, and dispose of it. Wash the hands after the waste is treated or discarded.

- Consider checking with the local health department for advice on waste disposal, if applicable.

6.4 What Really Works to Keep the Coronavirus Away?

The World Health Organization has announced that COVID-19, the new coronavirus-caused disease, has a higher fatality rate than flu. Nine deaths have been reported in the USA as of March 4, 2020.

Brian Labus, a public health professor, provides you with essential safety information, from disinfectants to food storage to supplies.

1. what should I do to stop being infected?

If people with a respiratory condition such as COVID-19 are sick, they cough or sneeze contaminants into the air. The virus could easily land on your head, nose, or mouth if someone is coughing near you. Such particles move just about six feet (1.8 meters) and fall very rapidly out of the air.

They do land on surfaces that you always hit, such as railings, doorknobs, elevator buttons, or subway poles, however. The average person often touches their face 23 times an hour, and about half of these touches are to the eyes, mouth, and nose, the mucous surfaces contaminated by the COVID-19 virus.

We public health professionals can't emphasize this enough: The safest thing you can do to protect yourself from a variety of diseases like COVID-19 is regular hand-washing. Although hand washing is preferred, hand sanitizers with an alcohol content of at least 60 percent can be an efficient alternative to using soap and water at all times, but only if your hands are not clearly soiled.

2. Wouldn't surfaces just be easier to clean?

Not exactly. Public health experts don't completely understand the role such surfaces play in disease transmission, so a virus that landed directly on you might still infect you.

We still don't know how long the coronavirus that causes COVID-19 can live on hard surfaces, while other coronaviruses can live on hard surfaces such as stair-railings for up to nine days.

Frequent cleaning will remove the virus if a sick person has infected a surface, such as when someone is sick in your home. It is important to use a disinfectant in these situations, which is thought to be productive against the COVID-19 virus.

Although particular products have not yet been tested against COVID-19 coronavirus, many products are effective against the coronavirus family at large. Cleaning recommendations using "natural" products such as vinegar are common on social media, but there is no evidence of them being effective against coronavirus.

You will need to correctly use these items according to the instructions, which usually means keeping the surface wet with the liquid for a period of time, often several minutes. Wiping the surface with a product simply isn't usually enough to kill the virus.

In short, any surface you reach in your day can't be properly washed, so hand-washing is always your best protection against COVID-19.

3. What about wearing masks?

Although people have turned to masks as protection from COVID-19, the masks also provide the wearer with nothing but a false sense of security. The masks that were widely available in pharmacies, big-box stores, and home improvement stores—until they were all purchased by a worried public—work well in filtering out large particles like dust.

The problem is that the particles carrying the COVID-19 virus are tiny and pass easily right through the masks of dust and surgical masks. Such masks that give some people with protection if you wear one when you're sick—like coughing into a tissue—but they can do nothing to protect you from other sick people.

N95 masks, which filter out 95% of the small virus-containing particles, are worn in health care settings to protect doctors and nurses from respiratory disease exposure. Such masks only offer protection when properly worn.

They require special testing to ensure they provide a seal around your face, and air does not flow in the sides, defeating the mask's purpose. People wearing the mask must also take extra precautions before removing the mask to ensure they do not contaminate themselves with the infectious particles filtered out by the mask.

If you don't wear the mask properly, don't remove it properly or put it in your pocket and reuse it later, you won't get any good even from the best mask.

4. Should I stockpile food and supplies?

In case of emergencies, you should have a three-day supply of food and water as a general preparedness measure. This helps to protect the water supply from shortages or during power outages.

While this is excellent advice for general planning, it does not protect you during an outbreak of disease. There is no reason to perceive COVID-19 to do the same disruption to our infrastructure that we Americans will see after an earthquake, hurricane, or tornado, and you shouldn't prepare for it the same way.

There's no need to buy 50 packages if you don't want to run out of toilet paper.

A quarantine of the form Wuhan is highly unlikely because a quarantine does not stop the spread of a disease that has been detected worldwide. Small disruptions in your daily life are the kinds of changes you can prepare for.

If you or component of your family gets sick, you will have a plan so you can't go out of the house for a couple of days. This involves stocking up on important things like food and medications that you need to take care of yourself.

If you get ill, the last thing you want to do is rush to the grocery store, where you'd show your disease to other people. Before asking for a refill, you should not wait until you are out of an essential prescription just in case your pharmacy shuts for a couple of days because all of its workers are sick.

You will also consider how to deal with problems such as temporary school or daycare closures. You don't need to plan something extreme; if you or your loved ones are sick, a little common-sense planning will go a long way towards making your life simpler.

6.5 Right mask to use for coronavirus

The industry offers a range of masks. Choosing a random mask doesn't secure you from COVID-19 infection. A standard medical mask, for instance, is not enough to protect you when used alone against the new coronavirus.

The department for Disease Control and Prevention (CDC) recommends a face mask with the N95 respirator to shield one from COVID-19 contracting. A face mask with an N95 respirator eliminates contaminants breathed into it from the skin. Such breathers filter out at least 95 percent of very tiny particles (0.3 microns). Such masks will filter out all kinds of organisms, including bacteria and viruses.

Why N95 respirators vary from N95 respirators, face masks, decreases the exposure of the wearer to airborne particles, from small particle aerosols to broad droplets. "If properly designed and used, minimal leakage occurs as the patient inhales air across the edges of the respirator. This ensures that nearly all air is guided into the filter media," the CDC states. Many face masks, unlike N95, are loose-fitting and provide only barrier protection against droplets, including large breathable particles. They will not efficiently filter out tiny particles from the air and do not avoid leakage when the user inhales around the edge of the mask.

When to use a mask:

You have to wear a mask if you are caring for a person with the suspected 2019-nCoV infection if you are safe. If you're coughing or sneezing, wear a mask. Masks are only adequate when used in conjunction with daily handwashing with hand rubbing based on alcohol, or soap and water. If you're wearing a mask, so you need to learn how to use it and properly dispose of it.

How to put on, use, take off and dispose of a mask

Wash hands with alcohol-based hand rubbing, or soap and water before putting on a mask. Cover with a mask, the mouth, and nose and make sure there are no holes between the face and the mask. Do not touch the mask while using it. When you do, clean your hands with hand rubbing or soap and water, depending on the alcohol. Replace the mask with a fresh one when humid and do not reuse single-use masks. Remove the mask from behind (do not reach the mask front) and dump it in a closed bin immediately. Then clean hands with hand washing, or soap, and water-dependent on alcohol.

Other preventive measures:

The CDC is suggesting routine preventive steps to avoid coronavirus spread. It involves avoiding sick people, not rubbing your eyes or nose, and covering up a tissue with your cough or sneeze. It advises staying at home and not going into crowded public areas or visiting patients in hospitals for people who are sick.

Chapter 7: How to prepare and take action to protect yourself and your family?

Despite urgent containment efforts, it has become evident that SARS-CoV-2 and the disease it causes-COVID-19-have spread throughout the globe to almost every country. While some countries report small numbers so far, it is almost certain that all will eventually see a nearby outbreak.

The question on the lips of all is the same: "What can I do to protect myself against coronavirus?"

The unfortunate answer is, not so much. Given the overwhelming amount of people advising you to improve your immune system with everything from vitamin C injections to sex, the only thing likely to affect the risk of COVID-19 is to wash your hands and try not to touch your face.

Nevertheless, what you can do, what anyone can do is help protect society. This doesn't mean that you personally won't catch the disease, although it may somewhat reduce your risk, but what it really means is that fewer people will get sick, and they'll be better cared for when they do.

So, you can do some practical stuff here to help reduce the spread of the latest coronavirus.

7.1 Some important tips
Social distancing:
Social distancing is a fairly easy concept-all the time. We come into close connection with other people. Hugs, hugs, a stranger's brief warm breath on your neck during your morning commute.

Seek to remain at a distance while practicing social distance, rather than getting in range. Using a way cooler fist bump, instead of kissing an acquaintance. Replace hugs with air-fives (remember for the quality of sound). Should not move in on the train people if you can stop it*.

All little stuff, but they can significantly influence how the virus spreads. That, in effect, will greatly shift the essence of the outbreak from a tragedy to something that can be treated far more easily.

Prepare at work/school:
One of the great things about infectious diseases is that they best spread when there are plenty of people around. This

includes in particular schools and workplaces, where children and adults are pushed together into tight, sweaty spaces.

So, plan to take measures to lower the risk of disease spreading. Schools will close, but not all of them, certainly, and not forever. By making sure children stay at home when they're sick, and enforcing simple rules like washing hands on a schedule during school hours, you can lower the risk to society.

Workplaces are a common story-if you're an employee, prepare to get home from work. Maybe you don't have to, but it's a good idea. If you're a manager, then be honest-sooner or later, people will get sick. Make sure that workers have sick time and find ways to keep working if employees have to stay home for a while. Start having what meetings you can by size, and try to make sure they are not crowded into small spaces without ventilation when people get together.

Practice at home:

Early knowledge from China shows one-way coronavirus spreads is through the household. Now, given the ambitious hopes of many millennials, it's obviously difficult to be completely disconnected from your family, but there are things you can do to help prevent the virus from spreading among your peers, relatives, and odd roommates.

If you become ill, isolate yourself from the family. When caring for sick loved ones, take precautions. More often, wipe off shared surfaces. Seek not to let your kids stick their hands too much straight into your mouth.

Wash your mouth, don't touch your face gently, sneeze, and cough. You know that I have already said this. Yet, again, it is worth telling. Washing your hands, not touching your face, and avoiding everyone else's coughing are some of the main ways you can help lower your risk of infection and protect everyone else.

Stay safe:
Overall, don't worry, but don't neglect the news entirely, either. There are some practical, clear steps we can all take in the coming weeks to help reduce the pressure on health services.

Recalling not to blame people when they spread the disease is important as well. There's no magic bullet against these viruses, and only the best precautions will the risk. Assigning blame may sound good, but eventually, it will only take control of infection more difficult as people seek to conceal their symptoms from the ravening masses.

7.2 Plans and preparations for coronavirus

Create a household plan of action:

Speak to the people you need to incorporate into your strategy.

Meet with members of your family, other relatives, and friends to discuss what to do if a COVID-19 outbreak happens in your group and what each person's needs will be.

Prepare strategies to care for those who may be at elevated risk for significant complications.

There is insufficient information about who may be at risk for serious COVID-19 disease complications. Using the evidence available for COVID-19 patients and evidence for associated coronaviruses such as SARS-CoV and MERS-CoV, older adults and individuals with ongoing chronic medical conditions may be at risk for more serious complications. Early results indicate that older people are more vulnerable to severe COVID-19 disease. If you or members of your family are at elevated risk of COVID-19 complications, please check with your health care professional for more knowledge on monitoring your health for COVID-19 possible symptoms. If a COVID-19 outbreak happens in your culture, CDC will prescribe measures to help protect people at high risk for safe complications.

Get to know your neighbors. Talk about Emergency Planning with your neighbors. If you have a website or social media page in your neighborhood, consider joining it to retain access to neighbors, information, and resources.

Identify aid organizations in your community. Make a list of local agencies that can be contacted by you and your household if you need an approach to information, health care services, assistance, and resources. Try having organizations that offer therapy or mental health services, lodging, and other resources.

Create an emergency contact list. Ensure your household has an up-to-date list of emergency contacts for family, friends, neighbors, carpool drivers, health care providers, teachers, employers, local departments of public health, and other community resources.

Learn healthy personal health habits and schedule plans at home:
take routine preventive steps now:
Remind everybody in your household of the importance of routine preventive measures that can help prevent the spread of respiratory diseases:

- Avoid close contact with sick people.
- Stay home while you're sick, except to seek medical treatment.
- Cover your cough with tissue and sneezes.
- Use standard household detergent and water to clean regularly touched surfaces and items every

day (e.g. tables, countertops, light switches, door buttons, and cabinet handles).

- If the surfaces are filthy, use a detergent and water to clean them before disinfection. For disinfection, a list of products with claims of emerging viral pathogens licensed by the Environmental Protection Agency (EPA), maintained by the American Chemistry Council Center for Biocide Chemistries (CBC), is available on the Novel Coronavirus (COVID-19) Combat Products icon outside. Always follow the manufacturer's directions for all items for cleaning and disinfection.

- Wash your hands daily with soap and water for at least 20 seconds, most importantly after going to the washroom; before eating; and after blowing your nose, coughing and sneezing; when soap and water are not readily accessible, use a hand sanitizer that contains at least 60 percent alcohol. Also, when your hands are obviously filthy, wash your hands with soap and water.

• Select a room in your house that can be used to distinguish sick family members from healthier ones. Identify, if possible, a separate bathroom for the sick person to use. If anyone gets sick, prepare to clean those rooms as needed.

Evaluate the effectiveness of an action plan for your household:

Discuss and note the lessons learned. Have your COVID-19 preparatory activities been successful at home, in school, and at work? Speak about the challenges and possible solutions contained in your strategy. Identify additional services which you and your household need.

Participate in community discussions about emergency planning. Maintain channels of contact with your peers (e.g., social media and mailing lists). Foster the value of practicing healthy personal health habits.

Continue to practice everyday preventive actions. Stay home when you're sick; cover with a tissue your coughs and sneezes; wash your hands regularly with soap and water; and clean your often-touched surfaces and items every day.

Care for the mental wellbeing of your family members.

Allow time to unwind and know that intense emotions are going to disappear. Take breaks from watching, reading, or listening to COVID-19 news stories. Connect with friends and relatives. Share your thoughts with others, and how you feel with them.

Help your child/children cope after the outbreak. Provide ways for kids to chat about what they have been through or what they think about it. Invite them to express their thoughts and ask questions. Since parents, teachers, and other adults see children in various circumstances, it is crucial that they work together to exchange information about how each child copes after the outbreak.

Checklist to get your households ready:

PLAN AND PREPARE:

Get up-to-date information from public health authorities about local COVID-19 events. Establishing an action plan for households.

- Consider household members who may be at higher risk, such as older adults and people with serious chronic diseases.
- Ask your neighbors what is included in their program.
- Create a list of local organizations that you and your household can contact if you need information, healthcare, support, and resources.
- Create a crisis contact list including family, friends, carpool drivers, neighbors, health care providers, teachers, employers, local department of public health, and other community resources;

- Select a room that can be used to isolate sick family members from those in your household.

Take everyday preventive actions:
- Regularly wash your hands
- Avoid touching the eyes, nose, and mouth.
- If you're sick, stay home.
- Use a tissue to cover the cough or sneeze, then chuck the tissue in the garbage.
- Disinfect and clean items and surfaces that are regularly touched Be vigilant if your child's school or childcare facility is temporarily dismissed or if potential changes arise in your workplace.
- Administrative controls refer to employer-work practices.

Protect yourself and others in case of an outbreak in your community:
- Stay home and talk to your healthcare provider if you develop cough, fever or shortness of breath
- If you develop cautionary warning signs for COVID-19, get medical attention immediately.

Emergency warning signs in adults:
- Breathing trouble or shortness of breath
- Constant chest pain or discomfort

- New distress or failure to arouse
- Bluish lips or face
- This list is not all-inclusive. For any other symptoms that are serious or of concern, please contact your healthcare provider.
- Keep away from sick people
- Limit close contact with loved ones as much as possible (about 6 feet)

Put your household plan into action:
- Continue daily preventive actions
- If someone in the house is sick, separate them into the separate room
- If you are caring for a member of the household, follow the recommended precautions and monitor your own health
- Keep disinfected surfaces
- Consider staying at home and away from the crew if you or a family member are an elderly person or have ongoing health issues
- Ensure that you have access to several weeks of medications and supplies in case you need to stay at home
- Keep distance from others who are sick and avoid direct contact with others

- Practice proper hand hygiene Taking precautions to better protect yourself.

7.3 How to prepare and stock up for a coronavirus quarantine

What's the difference between isolation, quarantine, and social distancing?

Social distancing, isolation, and quarantine each have specific goals, but both of these strategies are intended to restrict the spread of COVID-19, the disease arising from the novel coronavirus, and other communicable diseases.

According to the US Department of Human and Health Services, and the CDC, here is what each word means:

· **Social distancing**: Social distancing is used to restrict human interactions. You can see this happening with conferences being canceled, meetings being restricted, and schools being shut down. Individuals can often prefer to isolate themselves by avoiding public transport or opting for remote jobs. Many patterns of social distance include avoiding handshakes and standing more than three feet from other men.

· **Quarantine**: Getting quarantined (or self-quarantined) is when a person who is well— not ill or symptomatic— excludes or dramatically restricts their movement. This is used when a person has come into contact with an infected person (or is suspected of having done so) and must control their symptoms. Quarantine is often used for persons at high risk of COVID-19 contracting and having to limit their access to potentially ill people.

·   **Isolation:** Isolation is used when someone who is sick or exhibits suspected coronavirus signs is isolated from others who are well to help prevent COVID-19 from spreading. Patients can be isolated in a hospital in some cases, while those with manageable symptoms are isolated at home.

Who should follow these protocols?

Many US cities, including Seattle and San Francisco, are already practicing protocols of social isolation by canceling group activities and, in some cases, closing schools. Such cities declared the coronavirus epidemic to be a public health emergency, allowing health officials to enforce public-protection measures. For example, on March 11, health officials in California extended the policy for gatherings and ordered to cancel or postpone all activities with 250 or more participants.

But the question many people ask is: Should I take self-quarantine to avoid coronavirus exposure?

Currently, the CDC recommends all people over the age of 60 and the immunocompromised to maintain strict social isolation and also suggests that they "keep home as often as possible," but does not prescribe a full-on self-quarantine to avoid the disease from spreading or transmitting. Still, immunocompromised individuals (like me) can opt to self-quarantine or practice some form of combination of social distancing and quarantine as the virus takes hold in their societies.

Nonetheless, if the virus is as widespread as some medical experts expect, we may all find ourselves in some form of a quarantine (like the current protocol in Italy) or intense social isolation.

How to prepare for a coronavirus quarantine?

Preparing for a coronavirus quarantine has a lot more to do than hoarding toilet paper and bottled water. A quarantine guide should get you and your family ready to spend a lot of time at home, based on the recommendations of the CDC, HHS, World Health Organization, and CNET experts.

Remember that we don't have exact amounts— it may vary depending on your family size. Quantities may also be determined by how much time you want to be prepared for the quarantine (two weeks is a reasonable minimum, but one month is better).

Finally, remember that hoarding and planning are two very different things— we aren't recommending emptying the toilet paper shelves and those tasty little pot stickers from Costco. The advice is to procure enough supplies required for a possible quarantine.

1. Get a flu shot

This must be said: If you or another family member did not get a flu shot and you are still safe, go get one. The flu shot does not protect people from contracting COVID-19, but in a few important respects, it does help.

Having a flu vaccine significantly decreases the risk of contracting flu, allowing fewer hospital visits, enabling health care providers to deal with COVID-19 patients (and other diseases). You can help the body's immune system remain healthy by preventing the flu, so it can fend off other communicable diseases, such as COVID-19.

Finally, having a flu shot is about empathy and community responsibility; by reducing your risk of having flu, you are especially helping those with compromised immune systems stay safe and protected from COVID-19 as much as possible.

2. Stock up on these items

Many of us who work an eight-hour working day spend so much time outside our homes, at least. And we rely on our employers or other businesses for essential things such as toilet paper and meals during that time.

Grab the correct quantity of these items, as outlined by Ready.gov, after you have calculated the amount of quarantine time you wish to plan for. This is certainly not an exhaustive list— depending on the things that you rely on every day, your needs will differ.

Bath and hygiene

30 Day supply of drugs, including over-the-counter pain relievers, cough, cold and electrolyte medicine

- Toilet paper (which you'll use more while at home full time)
- Feminine hygiene products
- Hand soap (no, you don't even need a hand sanitizer)
- Washing detergent (ideally the powdered kind, which lasts longer)
- Slides, diapers, baby wipes, and other infant needs

Food and kitchen

There's no definitive selection of food products, but certain food items perform better than others. Even, you might want to check your kitchen toolkit, if you find yourself cooking more meals from scratch when trapped inside.

Pantry:
• Dry beans, rice and other grains, such as oatmeal
• Canned seafood, broth, and stews
• Essentials such as butter, salt, and pepper
• Smoothie mixtures and protein powder
• Coffee and tea
• Snacks that have a longer life shelf, such as dried fruit and nuts

Freezer:
• Meat and poultry (ideally vacuum-sealed), such as ham, beef, and pork
• Evite seafood that can spoil if not properly frozen
• Vegetables

Other:
  • Pet food (and treatments!)
  • All-purpose spray cleaning (here is the EPA list of COVID-19-fighting products)
  • A water filter (or filter replacement)
  • Dish soap and sponges
  • Paper towels

- Now is a very good time to get to know your Instant Bowl.

3. Get a better work-from-home setup

If you're lucky enough to keep working securely during the outbreak, you're going to want to confirm you have everything you need to work effectively. Justin Jaffe of CNET compiled this useful list of important work-from-home products, including a standing desk and guidelines for tracking. Find also some of the best practices, based on my knowledge of operating so far remotely:

Get ready and dressed for work each day. Doing this can bring you into a positive mindset, make you look presentable on video conferences, and keep some form of routine going.

Avoid housework. It's a rough one, but working from home doesn't mean doing the laundry, washing the dishes and sweeping up all day long. Make sure you clean before you start the day or before bed, to prevent some housework.

Coordinate meeting schedules. When you are quarantined with someone else who operates remotely, you'll want to arrange meetings so that you don't interfere. Only exchange schedules, or quickly communicate before the day begins.

Unless each one of you has an office or designated area, it does not apply to you.

Take breaks and stop working. Setting limits is the hardest thing to work from home. When you can rest, do an at-home workout or snack, make sure to plan breaks. Often make sure at some point that you are "clocking out" and putting away your laptop for the day. This will help you stay safe from home while you are working.

4. Change your routine

Losing your routine and being trapped indoors can put one's mental health under pressure. Here are some tentative items to look for.

Medical appointments:

When you need medical assistance, which does not require immediate admission, get to know telemedicine— or video appointment— services offered by your insurance company. My health company, for example, covers Doctor on Demand visits with a co-payment of $10. The physician may prescribe medications depending on the needs of you or your family members, which you might also prefer to have provided.

Exercise: You don't need to work out at home with a Peloton. Most YouTube channels deliver free fitness videos and exercise apps that give you an in-studio class experience on an equal footing. When you feel ambitious, you might even think about making a DIY Peloton. Here's our complete home workout guide.

Keep your spirits up. If the epidemic spreads and the death toll increases, many people will be extremely worried — or frightened even. Your mental health is equally critical during these periods, as is your physical health. Amanda Capritto talked to a psychotherapist who was providing some practical advice during the outbreak to remain alive.

What to do when you leave the house?
When you're taking part in a quarantine or self-quarantine that doesn't prohibit you from leaving home, there might be moments when you go out into the world, such as buying food or visiting a member of your family. Follow these suggestions when you do to prevent COVID-19 contamination, and make sure you wash your hands thoroughly and often. When you come home— or in case someone is visiting your home— make sure your house is sanitized. It means wiping commonly used surfaces like countertops, doorknobs, faucets, and tables using disinfecting materials. Many retailers are actually sold out with items to be disinfected, such as Lysol, online and in shops. So, here are some alternatives to sprays and wipes.

Chapter 8: Practical Advice to Protect Yourself and Your Family

The coronavirus is spreading worldwide, with more than 127,000 cases confirmed and nearly 5,000 deaths. At least 1,297 incidents and more than 35 deaths have occurred in the United States, according to a New York Times report.

"We suspect there will be thousands of more cases," Vice President Mike Pence said on NBC's "Today" show Thursday morning. His remarks came just hours after President Trump imposed immediate limits on most voyages from parts of Europe to the United States during a national television address.

Coronavirus is here, and it is easy to spread. The most vulnerable to the virus and its community damage are elderly people, those with underlying health problems, and those without a social safety net.

Even though life is sharply off-kilter as we know it, there are measures that you can take.

Most significant: Don't panic. You can help reduce your risk, brace your family, and do your part to protect others, with a clear head and some easy tips.

If you feel sick, stay home

Even if you don't have underlying health problems, be extra careful and protect others if you don't feel well. If the alert from Chancellor Angela Merkel is right, two in three Germans may become infected. This may well be a warning for the rest of the planet because the number of reported cases keeps increasing rapidly.

Many people who contract coronavirus do not get severely ill, though. If you have coronavirus, you'll probably just feel like you've got the flu. Yet holding a stiff upper lip isn't just foolhardy. It may also put others around you at risk.

So, stay home if you have a fever and a dry cough followed by exhaustion and shortness of breath. Don't go to work; don't carry your kids to school; don't go to the store; don't use public transport.

If you get permission to work from home, do. If your company usually does not provide sick leave, inform them of the danger or give them the CDC and Prevention guidance program.

Call your doctor if you experience high fever, shortness of breath, or another, more severe symptom. b (Coronavirus testing is also unreliable— there are not enough tests, so it is risky to go to a doctor's office and risk infecting others.) Instead, check the C.D.C. website and the local health department for guidance on how and when to be tested.

Wash your hands with soap. Then wash them again.
Wash the hands, wash the hands, wash the hands. The underwater splash flick isn't going to make it anymore.
A refresher: Wash your hands and clean with soap, taking care to get between your fingers and under your fingernails. Wash (or around the time it takes to sing "Happy Birthday" twice) for at least 20 seconds, and dry. The C.D.C. also advises that you stop rubbing your eyes, nose, and mouth with unwashed hands (which we know is difficult one).

Hand sanitizers based on alcohol, which should be rubbed in for about 20 seconds, can also work, but the gel must contain at least 60 percent alcohol. (No, homemade vodka from Tito doesn't work.) Also, clean "high-touch" surfaces, such as phones, tablets, and handles. Apple advises using alcohol with

70 percent isopropyl, cleaning gently. "Don't use the bleach," the company said.

The C.D.C. suggests wearing disposable gloves and washing hands thoroughly immediately after the gloves are removed to disinfect every surface. Some disinfectants in households licensed with the Environmental Protection Agency should operate.

Try to separate yourself from others, particularly if they seem sick, instead of shaking hands, wave, bow, or give an elbow knock.

Stay informed

Understanding what's true will protect you and your family. A lot of knowledge goes around, and understanding what's is happening can go a long way to protecting your family.

First of all, double-check. Mr. Trump misrepresented the travel restrictions in his address Wednesday night, and two crucial points were later clarified by the Homeland security and the White House.

A prohibition on European visitors does not extend to American citizens or to legal permanent residents. This will also have no effect on products shipped to the U.S.

The C.D.C. has up-to-date statistics and is a great resource for questions from the local health department.

With children, keep calm, carry on and get the flu shot
The good news is that cases of coronavirus have been very rare in children. Right now, there is no need for parents to worry, experts say; cases of coronavirus have been very rare in children.

The flu vaccine is a must because vaccinating children is the best defense against bacterial pneumonia for older people. And take the same safety precautions as you would take during a regular flu season: Promote frequent hand-washing, step away from people who appear sick, and get the flu.

Like with airplanes, making sure the metaphorical oxygen mask is on before helping others. Before addressing an outbreak with your children, make sure you first evaluate their knowledge of the virus and express your own anxiety. It's crucial that you don't cast off their fears and talk to them at an acceptable age level.

Be sure to connect with the school in your kid, including early dismissals or potential online instruction. Be prepared to close schools; several districts and universities around the globe have already taken that measure. In New York City, though, officials agree it's going to be a "last resort." It's also recommended that you talk with your boss about the child-care issues you have. If your kids are trapped at home, play some sports, put on a movie and try to make it feel like a holiday, at least for the first few days.

Don't stockpile masks

Face masks do not help. Face masks have become a sign of coronavirus unless you're already sick, but stockpiling them might do more damage than good.

Firstly, they're not doing a lot to protect you. Most surgical masks are overly loose to avoid virus inhalation.

(Masks can help avoid the spread of a virus if you are infected with it. The most powerful are the so-called N95 masks that cover 95 percent of very small particles.) Third, health care professionals and others who care for ill people are at the forefront. Last month, the surgeon general urged the public to avoid stockpiling masks, warning that the amount of resources available to physicians, nurses, and emergency personnel could be reduced.

But do stock up on groceries, medicine, and resources

Preparation is the most effective way to protect your family.

Stock up on a 30-day supply of food, household supplies, and medications, just in case.

Which doesn't mean you'll just have to eat ramen and beans. Below are tips for filling a canteen with shelf-stable and tasty foods. (Don't forget the chocolate.) Go to the pharmacy earlier rather than later if you are taking prescription drugs or are low on any over-the-counter product.

Then, in no specific order, if you have small ones, make sure that you're equipped with soap, toiletries, laundry detergent, toilet paper then diapers.

Are you concerned about the stock market? Take a deep breath.

With financial markets on a roller coaster, financial columnist The Times Ron Lieber recommends remaining tight even with the recent declines. We' going to settle in the long term. After all, "stocks are how your investments battle inflation, the market isn't an absolute instrument for your personal finances, and you're playing a long game." However, the volatility is real: stocks plunged on Thursday, and trading in the United States stopped minutes after the open. A seven percent drop in the

S&P 500 caused a so-called circuit breaker, a 15-minute pause intended to avoid crashing of markets.

"Take a second and take a deep breath," wrote Monday, Mr. Lieber. "So, ask a question: Did your long-term priorities change today? If not, your investments still have no need to adjust either.

Chapter 9: Traveling

The rapid dissemination of COVID-19 across the globe has caused confusion for the international travel industry.

Rising numbers of travelers chose to stay at home in fear of exposure to the latest coronavirus, which since late December, has spread to 79 countries, claiming more than 3,000 lives and infecting more than 100,000 people worldwide.

The virus, first identified in China's Wuhan and for which there is no vaccine yet, has triggered worries across the world, with governments closing borders with affected countries and banning entry or subjection to lengthy quarantines for travelers from outbreak areas. This is despite warnings against these travel limitations from the World Health Organization (WHO).

Organizations are calling off big events amid the disturbances, and main competitions are being canceled, postponed, or moved by global sports bodies.
Meanwhile, foreign airlines continue to suspend flights to hard-hit regions, including China, Italy, South Korea, and Iran.

Should you need to cancel or delay your travel plans?
It depends, say, public health experts.

The WHO advises that elderly travelers and those with underlying health problems postpone or prevent travel to areas experiencing ongoing COVID-19 transmission. This is because the disease can be fatal for people over 65 years of age or who have chronic illnesses, although mild in some 80 percent of cases.

WHO updates the list of countries experiencing local transmissions on a regular basis? Today it includes China, South Korea, Japan, Singapore, Malaysia, Vietnam, Thailand, and Indonesia in Asia.

The list includes Italy, France, Germany, Spain, UK, Switzerland, Norway, the Netherlands, Sweden, Croatia, Denmark, Finland, Greece, and Romania. Three Middle East countries-Iran, the United Arab Emirates, and Lebanon-are listed here.

Transmission of COVID-19 infections has been recorded in the Americas, the USA, Canada, and Ecuador.

Only one African nation-Algeria-is on the list.

Australia is on the list in the Oceania region, too.

Governments have issued their own travel alerts, too. However, this varies from one nation to another.

The British Foreign Office, for example, recommends against all but necessary journeys to mainland China, two cities in

South Korea, and the 11 cities in northern Italy that have been put under lockdown. The U.S. travel advice is broader, with travelers being advised to avoid China, South Korea, Italy, and Iran as a whole unless it is absolutely required.

What are other factors at play?

According to Crystal Watson, senior scholar at the Johns Hopkins Center for Health Protection, if you ever want to fly, even if your destinations are without major cases, you should understand your own risk factors and the feature of healthcare available in the area you would be going to if you become sick.

The thing to carry in mind is that if an epidemic happens, you might be put into quarantine.

"Travelers should be mindful that this is a possibility, they may be trapped for an extended period of time somewhere so they should prepare for that," she says.

"If this virus is recognized in more nations, travel bans may have less effect, so that countries can eventually avoid enforcing them. However, I think there will be considerably more disruption in the weeks and months ahead." Travelers will also find themselves with less incentive to venture away from home, as numerous tourist attractions have been shut down, and major events have been estimated.

Tourists are now being urged by the US Center for disease control and Protection to rethink all cruise trips to or within Asia, saying travelers are at an elevated risk of catching COVID-19.

This follows a major coronavirus epidemic on the cruise ship Diamond Princess, which was quarantined off Japan's Yokohama coast in February. There have been records of at least 706 infections and six deaths on the ship, causing some countries to turn away other cruise liners, including those without confirmed cases.

Another thing to remember is that your flight plans could get interrupted as other passengers choose not to fly. Also, in areas that have no significant incidents, large numbers of customer cancelations and no-shows have caused airlines to ground flights and disturb tens of thousands of passengers.

The International Air Transport Association (IATA) said several airlines are recording "no-shows" of 50 percent across a range of markets, and traffic has plunged on main Asian routes.

Meanwhile, you can have to bear the expense if you want to cancel a current flight or hotel booking. But, according to Jonathan Smith, a spokesperson for ABTA, a British trade group representing travel agencies and tour operators, more and more travel companies are showing versatility.

"When they plan not to travel, they will be bound by the usual terms and conditions of their travel company, and that would mean cancelation fees in most cases," says Smith. "What we find is that travel carriers have demonstrated a degree of flexibility with their customers in some situations, and our recommendation will be for passengers to speak to their travel company." A small number of airlines, such as JetBlue and American airlines in the United States, have agreed to waive cancelation fees on new reservations.

Travel insurance review site TravelInsurance.com recommends that those who book flights opt for the more costly "cancel for any excuse" policy because regular travel insurance does not cover traveler cancellations due to concerns about the destination, which involves coronavirus spreading.

How can you protect yourself if traveling?

When the travel plans go ahead, the WHO has a number of guidelines to reduce the risk of infection. They include regular washing of hands, covering nose and mouth while coughing, and avoiding close contact with people with symptoms.

When you have signs of illness while traveling, such as fever, coughing, or trouble breathing, the WHO suggests that you contact your nearest health care provider by telephone to inform them about your travel history.

Travelers can bring and use hand sanitizers on a regular basis while maintaining a distance of two meters from others where necessary, says Bharat Pankhania, a disease prevention expert at Exeter University in the UK.

"[One way to reduce the risk] is to know how the infection spreads and how it can defend itself, for instance, by maintaining its own personal safe zones when traveling," he says.

"The second is to equip themselves with knowledge from the country they are going to. Most people will see these two elements put together well shielded and secure."

Chapter 10: Hygiene environment

It's natural to look for ways to reverse the harm to our planet and keep our atmosphere safe with environmental problems, including water contamination, waste, and climate change. Many of us believe we're too small to make a difference, but we've seen the positive impact we can produce when enough of us is taking action.

What is control of infections, and why is it important?
All over there are germs like bacteria, viruses, and fungi! Some are actually helpful, as are those who live in our own bodies. Many are harmful and can cause severe illness or death.

The transmission of infections can take three main ways:
1. From person to person, directly
2. Indirectly through the supplies and equipment, and
3.Over the air.
The goal of control and prevention of infections is to prevent infection transmission and to keep both the senior citizens and their caregivers safe.

A Caretaker's Work Saves Lives:

As a caregiver, you have a significant role to play in disease prevention. The simple procedures for managing infections can literally save lives. Only way to do this is to make a safe and healthy climate.

Good people who have strong digestive systems are able to fend off germs. However, people who are elderly or unwell may have a greater risk of contracting illnesses and diseases. Older adults have a triple elevated risk of pneumonia and a twenty-fold higher risk of urinary tract infection relative to younger people.

Risks and Standard Procedures:

Many common risk factors for infection in older adults include:

- Malnutrition from unhealthily food or not eating enough food.
- Many medications that weaken the immune systems.
- Weakened immunity from diseases.
- Urinary catheter.
- Feeding tubes.
- Pressure ulcers.
- Long-term reduced mobility.

There are a few standard approaches you can use to minimize infectious disease spread. Those include:

- Hand hygiene, such as washing your hands.

- Cover your face and nose while sneezing or coughing.
- Having good personal hygiene.
- Make sure that you have good personal hygiene for the older adult.
- Using adequate food preparation and food storage.

Standard Precautions Maintain a Healthy Environment

Illnesses occur by blood or other body fluids.

Standard precautions are a series of guidelines designed to prevent disease transmission by blood and body fluid when treatment is given. The aim of these precautions is to protect you as caregivers. They are based on the idea that infectious germs may be present in all blood, body fluids, secretions, broken skin, and mucus.

It can transmit infectious diseases by blood or other body fluids. Common diseases transmitted by blood include hepatitis B, hepatitis C, and HIV.

Home caregivers are often colonized or contaminated with multi-drug-resistant species, or MDROs. The MDROs are bacteria and other germs that have developed antimicrobial resistance. One example of an MDRO is a form of Staph infection called Methicillin-resistant Staphylococcus Aureus (MRSA), which is immune to several antibiotics. If you have contact with the older person and their immediate environment, your hands and clothes will become contaminated.

If the older adult has been diagnosed with an MDRO or blood-borne infection, they and you as a caregiver will not know. This is why it is necessary to use the Standard Precautions to prevent such infectious diseases from spreading.

Importance of Handwashing in a Healthy Environment

The good thing is it's preventable to have such diseases. Hand-hygiene is the most effective means of avoiding germ transmission.

Every human being should be regarded as having an infectious disease. The elder caregiver should use: hand hygiene-explaining appropriate handwashing procedures Physical barriers-including gloves, robe, mask, eye protection, or face shield-to ensure a safe and secure atmosphere.

Properly disposing of laundry and hazardous waste-use towels only once after touching, and regularly wash the linen and when soiled.

Proper treatment of polluted areas and equipment-regularly and when soiled with body fluids disinfect the client's space.

If you are a licensed caregiver, all the adults you deal with should be given standard precautions to ensure a safe atmosphere. In certain situations, extra procedures, called "touch measures," will need to be used. Consult with your provider for specific instructions you need to follow with particular older adults.

When you are a family caregiver, ask the health care provider of your loved one whether you should take a particular set of measures, and inform the health care professionals who provide treatment for your loved one if the older person has an MDRO.

Tips to Prevent Infectious Diseases Spread Through the Air

Infectious diseases, such as a cough or sneeze, may also spread through the air. Such diseases include pneumonia and the common cold.

When caring for an older person with signs and symptoms of respiratory illness (such as fever, cough and/or sneezing) and whose health care provider has allowed them to stay at home, the following guidelines should be remembered: If possible, advise the older adult to cover the elbow rather than the hand with the mouth/nose.

When acceptable and necessary, put surgical masks on the individual coughing.

Holding hand hygiene after contact with respiratory secretions. Avoid direct touch (anything less than 3 feet) if necessary. That may not be easy to do when you are working. And if for whatever reason the older adult is reluctant to wear a mask, then you will.

If you have a respiratory infection: If possible, try to avoid direct contact with the older adult. If that is not feasible, wear a mask when caring.

Flu Vaccines Keep Adults Healthier

It's also critical that both the older adult and you, the caregiver, get your annual influenza vaccine, or "flu shot." Influenza is a harsh disease that can lead to hospitalization and, occasionally, even death. Each flu duration is different, and influenza infection may have different effects on individuals.

Even people who are healthy can get really sick from the flu and spread it to others. Older adults and individuals with chronic disease conditions are at especially high risk of experiencing complications associated with the flu.

About 80 and 90 percent of flu-related deaths have occurred in people 65 years of age and older during recent flu seasons.

"Flu season" will start as early as October in North America and last as late as May. During this time, in the U.S. population, flu viruses circulate at higher rates. An annual seasonal flu vaccine (either flu shot or nasal spray flu vaccine) is the safest way to lower the risk of having seasonal flu and spreading it to other people.

If more people get flu vaccinated, less flu can spread through the population. When you can get the flu shot as a caregiver, you can help keep the caregiver from catching the flu.

If, for whatever reason, you or the older person that you care for cannot get the flu shot, speak to your health care providers. While delivering treatment, you may need to wear a mask.

Ways to Keep your Environment Clean and Safe:

Refuse single-use items (especially plastic)
Straws, to-go cups, disposable razors, and reusable shopping bags are a few common examples in our culture of single-use

items. An easy way to make a major difference is to find a reusable alternative for these things that we use once and discard.

While it is possible to substitute all single-use goods with reusable goods, it can at first be daunting. To get started, at Minimal Domesticity, Lauren says to consider whether the product is going to be used for more than an hour.

If the useful life of the product is less than an hour, much like a plastic shopping bag, seek to substitute it with a more sustainable alternative.

Refuse them before they enter your life is one of the easiest ways to prevent such goods. And you're voting with your dollar. Unsubscribe from mailing lists and catalogs, carry your own pack, order a non-straw cocktail, and remove needless receipts.

Buy locally, eat more vegetables, and compost your food waste Focus on purchasing locally produced items while shopping, rather than imported goods. Locally purchasing means less shipping, less manufacturing, and less packaging. And when it comes to food—local means new... so much more tasteful.

Find and join the own CSA use Own Harvest. Or try your hand to grow from seeds with the ten easiest vegetables. When you are growing vegetables and fruit trees in your very own backyard, you do not have to make too many regular trips to the store.

Consuming more vegetables and fewer farm products tends to lower greenhouse gases. Seek to decrease your serving sizes and let the vegetables take center stage on your plate and in your belly if your family eats meat. And when you buy meat from a legitimate grass-fed source, be sure to do it without hormones or antibiotics.

Plant residues and kitchen waste make plants rich nutrient-filled food and manure, helping them grow faster. That process is called composting. You can now use it in your own home garden instead of throwing away food and wet waste for the plants.

Composting reduces the level at our landfills. Buried in a landfill, urban solid waste does not get enough oxygen and can contain methane. By comparison, a compost pile undergoes aerobic decomposition. Since it is exposed to oxygen, either by turning it or by using worms and other living organisms, it creates carbon dioxide rather than methane.

Plant trees and landscape with native plants

Green living areas are important to our towns and suburbs. The trees were stripped away by industrialization and suburban sprawl–our primary source of unadulterated oxygen. We are beautiful too, and they are doing their bit to keep our world safe. You will make the green space and unadulterated oxygen a possibility for our children by planting a tree today. The same goes for native plants for landscaping. Not only are they low maintenance, but they also save water, reduce carbon emissions, and encourage local wildlife protection. If you don't have your own lawn, donate a tree to voluntary organizations, such as the plant A Tree Foundation.

Green your transportation and travel habits

As much as 90 percent of U.S. road haulage relies on gasoline. You'll greatly reduce the carbon emissions when you walk, bike, or use public transportation. If you have to drive, make your orders at less busy times of the day, and you're not going to be trapped in traffic wasting gas. So, aim to simplify your outings- be mindful of how many different trips you can stop (this saves time as well).

Look for nearby destinations when it comes to vacations. Why don't you see all the lovely natural resources your own town or state has to offer? If you have to fly by air, consider purchasing carbon credits to offset the effect on the atmosphere.

Conserve water

Our access to new, clean drinking water is diminishing with companies pouring waste into our water sources. Examples of excessive water pollution include running taps, long showers, running the dishwasher half-full, and unregulated water leaks. Seek collecting rainwater in buckets or a tank of rain that can be used to water the plants in the yard, clean your cars, etc.

Reduce the use of chemicals & properly dispose of waste

Instead of purchasing disposable products such as plastic plates, spoons, and cups, opt instead for reusable, washable flatware. Get a bonus package at Goodwill. If they match, it doesn't matter. Bring your reusable travel mug into the coffee shop for on-the-go coffee lovers. In this small, easy act, you reduce the amount of garbage that you dispose of, and your coffee stays hot while you do so.

Most companies knowingly dispose of their own gasoline, paint, ammonia, and other chemicals. It is harmful for water and air because these toxic compounds are dissolved in the groundwater.

It's no wonder cancer rates have skyrocketed when all of these chemicals mix—help laws and corporate opportunities to keep our air safe. Agriculture is also a recognized polluter where chemical runoff is concerned. Even when it comes to your own gardens, stop over-fertilizing to ensure you don't lead to runoff algae blooms.

To keep your indoor air clean (and naturally fresh), and your waste less hazardous, use natural cleaning techniques, environmentally friendly paints, and recycled or non-toxic home improvement products.

Fall in love with Mother Nature

"Mounting research supports the belief that children [and adults] who frequently spend time playing and learning in the natural environment are happier, safer, wiser, more innovative and better problem solvers," shares Janice Swaisgood, National Coordinator of Nature Clubs for Families for the Children & Nature Network.

Essentially, we have to fall in love with nature if we want to be motivated to preserve our natural resources. Go out and wade in a stream, swim in a pool, walk on the beach or play on. Place your phone down, and go outside to see what kind of birds and butterflies flutter around your yard. You acquire an inherent vested interest or ownership in the natural world when you find a bird nest and watch (do not interfere) the hatching, rising, and finally flying away. It can take decades, even centuries, to strike a balance between ethical growth and a safe climate. But we can become better stewards of this world which we share as our home together.

Chapter 11: Coronavirus and business

As the Covid-19 crisis spreads through Europe and the US to new epicenters, companies are struggling to organize responses. Due to the unpredictability of dynamic diseases, a lack of appropriate professional knowledge, and the lack of plug-and-play guidance from government or foreign authorities, there are no simple answers.

Obviously, each local scenario is different, but we agree that there are incentives for businesses to learn from those in regions where the answer to the epidemic is weeks ahead.

According to our review of the high-frequency data on proxies for the movement of people and products, development, and trust, China appears to be in the early stages of an economic rebound. Although this recovery could be vulnerable if a new wave of local infections arose, several Chinese companies have already switched to recovery and post-recovery preparation beyond crisis response.

How Have Chinese Companies Responded to Coronavirus?
1. Look ahead and constantly reframe your efforts.

Crises, by definition, have a highly fluid trajectory, requiring a constant reframing of conceptual models and plans. Early confusion gives way to exploration and making sense, then crisis preparation and response, recovery strategy, post-recovery strategy, and, ultimately, contemplation and learning. This process must be quick— and thus led by the CEO— in order to avoid getting caught in complicated internal management processes and responding slowly to changing circumstances.

Many of the fastest-recovering businesses in China looked proactively ahead and expected these changes. For instance, Master Kong, a leading instant noodle and beverage manufacturer, assessed daily dynamics in the early stages of the outbreak, and periodically reprioritized efforts. This expected hoarding and warehousing and shifted its attention away from physical, broad retail outlets to O2O (online-to-offline), e-commerce, and smaller shops. Through actively monitoring the re-opening plans of retail stores, it was also able to change the supply chain very flexibly. As a result, the supply chain had improved by more than 50 percent only a few weeks after the outbreak, and it could supply 60 percent of the stores that were reopened during this period— three times as many as its rivals.

2. Using a bottom-up, adaptive approach to complement the top-down efforts.

Rapid, organized responses warrant leadership from the top down. But responding to rapid change often includes decentralized initiative-taking, with distinct complexities in different societies. Some Chinese companies combined the two strategies successfully, creating a top-down structure through which employees can be innovated.

For example, Huazhu, which operates six thousand hotels in 400 cities across China, formed a crisis task force that meet up daily to review procedures and provided top-down instructions for the entire chain. It also leveraged its internal information portal, an application called Huatong, to ensure that employees and franchisors were provided with timely information. This enabled franchisees to adapt central guidelines, in terms of disease conditions and local public health programs, to their own local circumstances.

3. Proactively create clarity and security for employees.

In a crisis, when the situation and the available knowledge continuously change, guided by the exponential mechanism of contagion, it's hard to find clarification. For practical purposes, official advice can be missing, inconsistent, out-of-date, or not granular enough. Additionally, a multitude of media outlets with varying viewpoints and suggestions intensify the uncertainty. Employees will continue to embrace new ways of operating, but unless they have straightforward, concise knowledge and general guidance, they will not be able to do so.

Some Chinese firms provided very constructive guidance and employee support. For example, China's largest manufacturer of kitchenware, Supor, has implemented very detailed organizational guidelines and procedures for its workers, such as directives for restricting exposure when dining in canteens and emergency measures for irregular circumstances. Moreover, from the early stages of the outbreak, the organization instituted safety checks for employees and their families and procured protective equipment. It was well prepared to resume work in time, reopening some production lines in the second week of February.

4. Reallocate labor flexibly to different activities.

Employees were unable to carry out their daily operations in hard-hit industries, such as restaurants. Many innovative Chinese firms, rather than furloughs or layoffs, deliberately reassigned workers to new and beneficial tasks, such as disaster planning, or even loaned them to other businesses.

For example, more than 40 restaurants, hotels, and cinema chains streamlined their staffing to free up a significant proportion of their workforce in response to a drastic fall in sales. They then shared those workers with Hema, a "modern retail" supermarket chain owned by Alibaba, which, due to the sudden rise in online sales, was in desperate need of work for delivery services. O2O teams, including Elle, Meituan, and JD's 7Fresh, also followed the lead by borrowing restaurant labor.

5. Shift your sales channel mix.

In affected regions, personal-to-person and brick-and-mortar retailing were severely limited. Agile Chinese companies quickly redeployed distribution activities to new platforms in both B2C and B2B businesses.

For example, during the crisis, cosmetics firm Lin Qingxuan was forced to close 40 percent of its stores, including all its Wuhan locations. Nonetheless, the company redeployed its 100+ beauty advisors from those stores to become online influencers who leveraged digital resources like WeChat to digitally connect consumers and boost online sales. As a result, its revenue in Wuhan increased by 200 percent compared to the revenue of the previous year.

6. Use social media to coordinate partners and employees.

With remote working and a new set of dynamic communication problems, many Chinese businesses have taken to social media sites to organize workers and partners, such as WeChat.

For instance, Cosmo Lady, China's largest underwear and lingerie company implemented a program to increase its sales through WeChat, enlisting employees to promote their social circles. The organization created a sales ranking for all employees (including both the chairman and the CEO), helping to inspire the rest of the employees to take part.

7. Prepare for a faster recovery than you expect.

Just six weeks after the initial outbreak, China appears to be on early recovery stages. Congestion delays are now at 73% in 2019, up from 62% in the worst part of the outbreak, suggesting that the flow of people and products is resuming. Similarly, coal consumption appears to be rising from a 43 percent low to currently 75 percent of 2019 rates, suggesting a resumption of some demand. And trust appears to be returning as seen in real estate transactions, which had dropped to 1% of 2019 rates but since then have fallen back to 47%.

Although it is hard to predict the extent and length of the economic effects in other countries, China's experience points to a scenario that companies will plan for. Considering the time, it takes to formulate, disseminate, and execute new strategies in big business, recovery preparation needs to begin when you're still responding to the crisis.

A luxury Chinese travel agency, for example, facing a collapse in its short-term market, reoriented around longer-term preparations. It allowed staff to use their resources to update internal processes, develop expertise, and create new products and services to be better equipped for eventual recovery, rather than the headcount.

8. Expect different recovery speeds for different sectors.

It comes as no surprise that markets and product classes are emerging at varying speeds, requiring distinct approaches. During the first two weeks that China's epidemic has intensified, but leading industries such as software and services and healthcare equipment and services have stabilized, stock prices have dropped across all industries within a few days and have since risen by an average of 12 percent. The majority of industries recovered more gradually but within a few weeks exceeded prior peaks. And the hardest-hit sectors— such as transportation, retail, and oil, comprising 28 percent of China's largest stock market capitalization— are still down by at least 5 percent and display only modest signs of recovery.

That means businesses need to calibrate their business-by-business approach — and big corporations need to calibrate their divisional approach. For example, a large global food &

beverage conglomerate used the crisis to intensify the long-term changes in its product mix in China (the second largest market for the business worldwide), including growing its emphasis on health-relevant goods, imported products, and online distribution platforms.

9. Look for opportunities amid adversity.

Although China's crisis has affected all industries to some degree, demand has risen in many different areas at a more granular level, which includes e-commerce B2C (especially door-to-door models), e-commerce B2B, remote meeting services, social media, products for grooming, health insurance, and other product categories. Some Chinese players were swift to mobilize to meet these needs.

For example, Kuaishou, a $28 billion social video site, supported online education services to counter the closing of schools and universities. The business and other video networks have collaborated with the Education Ministry to open a nationwide online cloud classroom to support students. And a major chain of restaurants leveraged downtime to prepare a new semi-finished dishes range, catching the increased need and demand for home cooking during the crisis.

10. Adapt your recovery strategy by location.

Regional public health strategies, disease dynamics, and administrative guidelines can establish recovery patterns that differ from location to location— perhaps not following the company's regional structure. That needs an approach that is versatile.

For example, a leading Chinese dairy company (a $10bn business with a large production base and deep domestic distribution in China) built a segmented strategy focused on dynamics of regional and city recovery, as well as its own supply chain infrastructure and salesforce density. In a staggered strategy, the expected supply from factories in highly affected areas was allocated to factories in other regions. Marketing, advertising, and budget allocation practices were also constantly updated to reflect regional variations in expected speed of recovery, customer mood, and needs.

11. Rapidly innovate around new needs.

Besides rebalancing the product portfolio, emerging consumer demands are also generating innovation opportunities. Most companies would concentrate on defensive measures when confronted by crisis, but some Chinese firms have confidently innovated around emerging opportunities.

The insurance industry is traditionally conservative, but Ant Financial has added free coronavirus coverage to its products in response to the crisis. The intervention met a need for a customer while fostering recognition of the online services of the business and enhancing consumer loyalty. Relative to the previous month, it expects a 30 percent rise in health insurance revenue in February.

12. Spot new consumption habits being formed.

Many trends are likely to continue after the recession, and other markets are likely to reemerge into new business realities in China and beyond. Nonetheless, the SARS crisis is also credited with speeding up e-commerce adoption in China. It's too quick to say for sure which new behaviors will last in the long run, but some strong possibilities include a move from offline to online education, a health care delivery transition, and a rise in digital B2B networks.

In the post-crisis world, several Chinese businesses are still preparing for those changes. For example, a global confectionery manufacturer's Chinese business has intensified its current efforts at digital transformation. The company scrapped offline promotions for Valentine's Day and other promotional events, instead reinvesting money into digital ads,

WeChat services, and collaborations with O2O networks to leverage new customer habits during and after the outbreak.

How to Survive and Keep Your Company Alive?

It is not the first recession in the world, and it will not be the last recession. It's not the first pandemic in the world, and it won't be the last one. The trick for entrepreneurs is to keep a cool head around you, don't do something foolish (for example, if you've never used weapons, now's not the time to buy one and start keeping it while wearing a gas mask on city streets) and take a war footing when you're running your company through choppy waters for 12 to 18 months.

Your business will have commitments that it is supposed to meet as the crisis continues. When the crisis decreases and the courts reopen, the business will have to include an accounting of its commitments and response for how it has, in the meantime, fallen short.

Here's how:

1. Protect your employees
In my mind, the early-stage founders ' first task is not to protect their creditors, but their workers.

Understanding that an officer of a corporation's formal legal obligation is to support the company's performance for the good of its owners, early-stage companies typically fall into one of two categories—founders-owned, or founders-and-VC-owned—and shareholders ' identities shift a lot on where the corporate interests of a company continue to reside.

In my experience, strictly founding businesses prefer to see their closest workers—who help the business make money—as assets, and see VCs as a diversion.

On the other hand, founding-and-VC-owned businesses tend to view their creditors and investor partnerships as the company's biggest asset, at least until they manage to get the business going under their own control. The interests of creditors in these companies continue to take precedence.

There is nothing wrong with either approach; sometimes, the software that you are developing is so early that if you want to spin up a company, you have no choice but to accept investor funds. Keep in mind, that (a) venture investment recognizes a high degree of loss as inevitable and (b) failure to keep your employees safe from an outbreak that result in the employee's illness or death, potential forward transfer to third parties and adverse health effects for you, your company and society at large.

The venture investors, put another way, can afford to lose some capital. Your workers can not afford to get ill. Now, not next week, but now is the time to write and prepare strategies to stop workers commuting, staggered off-peak traffic, changed paid sick leave and disability cover, and work from home.

Communicate such policies to the employees. Open e.g., The contingency plan given by Coinbase as an example of best practice. These measures result in a small reduction in the office's productivity or less "face time," but they will save lives and protect the employees, the people you'll have to deal with until the epidemic subsides face to face.

2. Cut your burn rate. Now.

Once the Saudis dumped the OPEC equivalent of a nuclear bomb on the markets, tanning the price of a Brent crude barrel to $30, it was apparent that the coronavirus collapse would have broader consequences for the U.S. economy–primarily the bankruptcies of many Middle American shale oil firms.

These companies will be among the victims of the recession ahead. Unless you don't want to be a statistic, you have to prepare for at least one year with very unfavorable market conditions.

Don't wait for the stuff to turn around or expect the markets will turn around; past globe-spanning outbreaks have taken 12 to 18 months to shake off fully and, in the absence of a pharmacological intervention that makes the coronavirus outbreak an unpredictable yet non-lethal disease, you can expect a very bumpy ride next year. Costly office room, dead weight on the staff-it all needs to go, now. Don't stop taking tough calls.

3. Whatever the deal is, close it. Now.

When there is an offer on the table—either an acquisition or a venture funding—on less than ideal but nevertheless reasonable terms, take it according to the above. Now is the time to commit an offense in closing any commercial transaction that will promote the short-term survival of your company or any return on capital for yourself or your investors.

The same applies to new customers being cut off. If you only have a period of 12 to 18 months, start working on sales-now. Investors in VCs are herd animals. Right now the herd is living out the conspiratorial prepper fantasy we have been entertaining for years in the tech crowd: hoarding freeze-dried food, buying crossbows (most VCs live in San Francisco or New York, so they can't own firearms) and planning to hole up in long haul bunkers or apartments in the Bay Area. Your company is not high on your list.

4. Review your insurance.

If you're looking for coronavirus insurance coverage and associated business interruptions, I've got bad news—there's probably no prospective cure here, and many markets that have insured this form of risk may go out of business.

This doesn't mean that you're not protected at all or that you shouldn't put any kind of cover in place. Unless you're a really early-stage organization, you're obviously going to want to put in place reasonable provisions for e.g., general liability that some of the leases and contracts would entail.

If you have insurance already in place, check your plans. Coverage can be found in unexpected ways-and counsel assistance will help you find it. If you succeed in uncovering a scheme that protects a coronavirus-related failure, make sure that you seek counsel before submitting it to maximize the probability of its effectiveness before filing a lawsuit.

5. Review and restructure your contracts.

Look at your supplier agreements, lease agreements, and other agreements as part of your burn rate analysis, which cost you a lot of money and without which you could do better. Unless there is a force majeure clause authorizing you to cancel the deal, determine that it would not be a bad idea to do so.

When the coronavirus has interfered with the contract in such a way that it has been practically made difficult to execute, common-law remedies such as annoyance or inability can also be invoked. Also, early termination requirements can be expressly in effect, which are explicitly applicable. If you know what your contract terms are, it will help you figure out which ones you should throw away.

And if you think you can't throw them free, it may be worth trying to restructure the contract by contacting your counterparties. If you don't ask for it, you won't be getting a lowering in your rent or released from a fixed-term deal. A mutually negotiated settlement in advance, after the event, is almost always preferable to acrimonious litigation.

6. Put in place a succession plan.

A CEO and an intern are almost the same in the eyes of a virus; however, if the CEO is older, the CEO is likely to be more susceptible to the virus than the more junior staff members.

Don't give the keys of the whole kingdom to one guy, like Quadriga. Have disaster management plans in place and a central chain of command so that the organization can continue to function as an ongoing concern if one staff member is taken ill or dies—backup the data across multiple geographic locations.

Chapter 12: Other Events that are affected

The coronavirus continues to wreak havoc in worldwide industries— from technology and sports to movies and politics. Several businesses have shut down factories and banned business-related travel; large cultural organizations such as the urban Museum of Art in New York have closed; political demonstrations have been canceled; and significant-tech industry conferences such as the Facebook's F8, E3 gaming show, the Google I/O, the Geneva Motor Show, and Mobile World Congress have been called off.

The NBA suspended the season on March 11, the same day the WHO called the outbreak a pandemic. Many cultural activities were postponed, such as the Coachella Valley Music & Arts Festival and the Miami Ultra Music Festival.

COVID-19, the latest coronavirus-caused outbreak, has killed more than 5,000 people and infected more than 137,000.

Here's how the epidemic impacts our lives:

Sporting events:

- On 11 March, the NBA suspended the rest of the 2019-2020 season after Utah Jazz player Rudy Gobert had reportedly tested for coronavirus positive.
- Major League Soccer postponed the season on 12 March as it' continues to determine the effects of COVID-19 with its public health officials and medical task force.'
- Many major NCAA Division I conference, including the SEC, Big 10, Big 12, ACC, and American Athletic Conference, announced on 12 March that their respective conference tournaments would be canceled next week. They have canceled the March Madness tournament.
- The NHL joined the list of leagues that were suspending their season on 12 March. The MLB has announced it has postponed spring training games and will delay the regular-season start of the 2020 baseball by at least two weeks.
- The Australian Grand Prix, the first Formula One season race in 2020, was canceled after a team member tested the virus for positive.
- Long Beach, California officials, have called for the cancelation of all major events through April, including the city's Grand Prix.

- The English Premier League postponed all matches until 3 April at least after a player and coach tested positive for the virus on different teams.
- On 13 March, the Augusta National Golf Club postponed the 2020 Masters Tournament by citing' the health and well-being of those connected with these events and the people of the Augusta region.'
- According to the Boston Athletic Association, the Boston Marathon will move its date from 20 April to 14 Sept.
- EA said its professional gaming series, including the Apex Legends Global Series, FIFA online series, EA SPORTS FIFA 20 Global Series, and Madden NFL 20 Championship Series, is suspending all live events. The suspension started on March 13, and will continue "until the global coronavirus situation improves." Online activities will begin, EA said, where participants and workers are separated.

Cultural happenings and institutions:

- The New York Urban Museum of Art announced it would officially close on March 12.
- The Coachella Valley, Music & Arts Festival, has been moved back to November, with the Miami Ultra Music Festival delayed to next year.
- The WonderCon Anaheim, set to be held in April, was postponed on 12 March.

- The Kentucky Derby Festival adjourned all activities until April 4.
- The Tribeca Film Festival, set to take place in April, was postponed after New York barred 500 or more participants from the events. Broadway theaters closed on 12 March.
- Beyond Wonderland SoCal, to be held in March was postponed until June. EDC Las Vegas is also due to take place in May.
- On March 18, Patreon will host a Livestream to benefit musicians impacted by canceled concerts and events.
- Billie Eilish said that she is postponing some dates of her North American tour "until further notice," adding that specifics of the rescheduled dates will be revealed early.
- A handful of late-night events in New York, including The Tonight Show Starring Jimmy Fallon and Seth Meyers Late Night, are set to cease production until 30 March at the earliest. The Ellen Show revealed on 13 March that production would be halted until 30 March.
- ABC reveal The Bachelorette has been delayed, Warner Bros, with Clare Crawley as the lead looking for love. TV Group revealed March 13.
- A number of high-profile movies, including A Quiet Place 2, Mulan, and the following James Bond sequel, No Time to Die, have seen their release dates pushed

back. Walt Disney Studios said on Friday that production on certain live-action movies is pausing.

- The Jonas Brothers reported that their residency in Las Vegas was canceled from 1 to 18 April.
- Kelly Clarkson put off her residency in Las Vegas from April 1 to July.
- Washington Monument is to be officially closed on 14 March, the National Park Service reported on 13 March.

Theme parks:

- In February, Disney briefly reopened its Shanghai and Hong Kong theme parks because of coronavirus. It's estimated the switch would cost about $175 million to the company.
- On 12 March, California Gov. Gavin Newsom and state health officials called for the cancelation or postponement of meetings with 250 or more participants by the end of March. The change does not extend to conditions such as attendance at the workplace, job, or necessary public transport. Newsom said it didn't include Disneyland in a press conference, but Disney later shared an announcement that the theme park was eventually going to close.

- Universal Hollywood Studios are also officially closing from 14 to 28 March. Universal CityWalk is expected to remain open.

Political events:
- Louisiana postponed its presidential primary on 13 March. It was originally expected to take place on April 4. This has now been scheduled for 20 June.
- Democratic presidential candidates Bernie Sanders and Joe bridan Sanders have canceled rallies in many states leading up to primary elections.

Facebook:
- Canceled the F8 developer conference, the biggest corporate event of the year in which CEO Mark Zuckerberg informs the world on Facebook's innovations and challenges, will then host local developers and online events meetings.
- Curtailed travel to China by employee.
- Canceled an early March marketing conference, which was supposed to attract 4,000 participants.
- Gives free WHO ads to provide health statistics.
- Expects development delays for the Oculus VR headset.

- Banned advertisements are offering a cure for coronavirus.
- The SXSW festival withdrew.
- Confirmed that a contractor had tested positive for coronavirus at its Seattle offices.
- Its Seattle office officially closed until 9 March, with workers being encouraged to operate at least from home until 31 March.
- Will continue paying seasonal employees who are unable to perform their jobs remotely.

Apple:
- On 13 March, the company said its Worldwide Developers Conference would only be online this year.
- Apple briefly shut down its 17 Italian stores as the nation locked off. While they have mentioned, the company had completely closed its 42 stores in mainland China.
- He said his quarterly profit guidance would be absent due to the impact of coronavirus.
- Forced to search for alternate sources for parts after suppliers closed in Wuhan due to the outbreak in that region.
- Retail stores allegedly warned that replacements for severely damaged iPhones would be in short supply.

- CEO Tim Cook allowed the majority of his company's global workforce to work from home. The business has recently restricted travel to Italy, China, and South Korea, and is heavily cleaning offices and shops.
- It has been confirmed that an employee of its campus in Cork, Ireland, has tested COVID-19 positively and isolated.
- Pulled out of SXSW festival officially.
- As of March 6, several Apple retailers in New York City reportedly had run out of iPhone 11 smartphones.

Google:
- Close all of its offices temporarily in mainland China, Hong Kong, and Taiwan.
- Commercial travel limited to China and Hong Kong.
- Told staff of immediate family members who have been returning from China to work at least 14 days from home.
- Keeping European offices open even after an employee was infected with coronavirus in Zurich.
- Canceled the Google News Initiative Summit in Sunnyvale, California, scheduled for late April.
- Converted its cloud conference annually, which last year attracted 30 000 attendees, to a digital-only gathering.

- Will continue paying seasonal employees who are unable to perform their jobs remotely.
- The annual I / O developer meeting, to be held May 12-14 in Mountain View, California, has been canceled.
- Google limits visitors to its New York City and San Francisco Bay Area offices, cancels all in-person job interviews and asks Japanese and Korean workers to work from home, Google announced on March 9.
- On March 10, Google launched a COVID-19 fund to cover paid sick leave for all temporary employees and vendors worldwide who have suspected symptoms of coronavirus or are unable to attend work because they are quarantined.
- Google reiterated its March 10 suggestion that all North American employees operate from home.

Microsoft:

- Announced that it is "recommended" to all employees in Seattle, Puget Sound and San Francisco Bay Area who are "in a work that can be done from home by March 25." Company president Brad Smith has said that they will continue to pay their daily salaries to their hourly campus staff even though their hours of service are shortened.

- It Warned investors that earlier estimates would likely miss revenue in the business segment, which contains its Windows operating system and Surface devices.
- On March 12, Microsoft announced it is canceling its Build 2020 developer event in person. The Build Show will proceed in a virtual fashion, officials said, in the same mid-May time slot that was scheduled for the regular meeting.

Twitter:
- On 11 March, all workers worldwide were allowed to operate from home.
- Retrieved from SXSW, where CEO Jack Dorsey would have delivered a keynote address.
- All non-critical corporate activity and workplace activities suspended.
- CEO Jack Dorsey initially intended to spend a few months in Africa in 2020 but said on March 5 that he was re-evaluating those plans "in the light of COVID-19."
- Twitter announced on March 6 that his office in Seattle had been closed for deep cleaning after an employee had been "advised by his doctor that they are likely to have COVID-19."
- Will continue paying seasonal employees who are unable to perform their jobs remotely.

Amazon:

- More than 1 million listings have been deleted for products claiming to cure or protect against coronavirus, according to a Reuters survey.
- Removed thousands of products for price gouging from merchants;
- The employee at his headquarters in Seattle has screened SARS-CoV-2 positively and is now in quarantine.
- Told Seattle area workers to work from home before the end of March, if necessary.
- Cancelation from the SXSW festival.

Dell:

- Dell officially told its 2020 tech conference attendees that it has switched to a "virtual environment" because of concerns about coronavirus. Keynotes and some meetings, the note states, will be available. Dell Technologies World was expected to go to Las Vegas from 4-7 May. Dell did not respond immediately to a request for comment.

Foxconn:

- In early February, its workers were told not to return to work at their Shenzhen, China offices until further notice.
- Plans to make its factories operating at full capacity by the end of March, Bloomberg reported.

Airbnb:

- Guests will be able to cancel reservations without penalty if they have booked through April 1 in China.
- Introduced a new system called "More Convenient Reservations," allowing travelers to cancel qualifying reservations without being charged and forcing hosts to refund the reservation irrespective of any prior cancelation policy. Travel vouchers will be refundable to Airbnb's service fees for trips booked through June 1.

Uber:

- Approximately 240 user accounts in Mexico were temporarily suspended to avoid coronavirus spread after those users had come into contact with two potentially virus-exposed drivers.

- Declared any driver or Uber Eats delivery person who has been diagnosed with COVID-19 or who is personally requested by a public health authority to remove himself will receive financial assistance for up to fourteen days while the account is on hold.
- Customers now have the choice, when ordering Uber Eats delivery, to leave a note in the Uber Eats app querying the delivery person to drop the food at the door, rather than making an in-person offer.
- They formed a support network to assist public health authorities in their response to this epidemic. The company said this team could temporarily suspend riders or drivers accounts which have been reported to have contracted or exposed to COVID-19.
- Workers strongly advised working from home in many countries where COVID-19 cases are growing, including the US, Canada, Japan, Europe, and South Korea. The advice lasts until 6 April.

Lyft:
- Allowed workers at its San Francisco headquarters to operate from home after one team member was found to be "in touch with someone who was exposed to COVID-19."

- Has partnered with EO Products to deliver more than 200,000 hand sanitizer bottles and other cleaning supplies to passengers. In mid-March, the company also said that it would "provide funds to drivers if they were infected with COVID-19 or placed by a public health agency under individual quarantine."

Tesla:
- The new plant in Shanghai was closed for a scheduled week and a half after the Chinese government had ordered private companies to temporarily stop operations.
- Investors warned that shutdown might have a "slight" effect on first-quarter earnings.

Nintendo:
- The development of its famous Switch smartphone in China was reportedly "having some effect from coronavirus."
- Its US branch is currently letting workers in California and the state of Washington work from home.

IBM:

• IBM tweeted March 9. Workers living and working in Westchester County or New York city are allowed to work from home until further notice if their job allows. All areas are susceptible to population dissemination of the coronavirus.

Salesforce:

• On March 5, Salesforce CEO Marc Benioff demanded that all workers in Seattle, Kirkland, and Bellevue, Washington, work from home for the entire March period.

Cloudflare:

- Cloudflare provides its Cloudflare for Teams, a suite of security software, free for six months to small businesses affected by a coronavirus. It's also helped create an industry initiative to promote small companies, called OpenforBusiness.org.
- The organization lets workers operate remotely in impacted areas.

Cisco:

- Cisco makes free use of its Webex communication and video calling software to states, health care providers, companies, educational institutions, and nonprofits.
- Security products such as Cisco Umbrella, Duo Security, and Cisco AnyConnect Secure Mobility Client are also provided by the telecommunications company to remote workers with free trials at no extra charge through July 1.

Discord:
- Discord eases the cap of its Go Live sharing service from 10 people at a time to 50, so teachers can take classes, communicate with colleagues and community meet remotely.
- It will continue for "as long as the need arises," CEO Jason Citron said in a blog post. He also cautioned that there is likely to be an increase in demand for the service, and it could suffer performance issues.

Tech industry events:
Thanks to fears about the coronavirus, numerous popular industry conferences have been postponed or redesigned. These include:

- E3, the year's biggest gaming convention expected to open in LA on June 9. Some exhibitors will also hold online events like Microsoft and Ubisoft.
- Mobile World Congress, an international technology conference scheduled to open in Barcelona on 24 February.
- March marketing summit for Facebook
- The Geneva Motor Show, one of the huge autos shows of the year since the Swiss government banned all activities involving 1000 or more participants.
- Adobe Annual Meeting in Las Vegas. Alternatively, the company says it will be selling some material online.
- Google I / O, the largest gathering of the year in which the tech giant is unveiling its newest products and initiatives.
- The GPU Technology Conference, usually held in San Jose, draws about 10,000 participants. Now it'll be a digital-only event with a webcast planned for March 24.
- Annual Partner Meeting on Snap. Snap, the Snapchat messaging app's parent company, said it'd be an online-only event, with a keynote planned for April 2.
- Live matches for Activision Blizzard's Call of Duty League set for 2020. These will be moved to online competitions only after all Overwatch League homestand matches have been canceled by the organization through April.

In addition, the annual Game Developers Conference, originally scheduled for March 16-20 in San Francisco, has been delayed to an unspecified date after exhibitors including Amazon, Microsoft, Epic Games, Sony, EA, and Facebook have dropped out.

The annual RSA Conference on cybersecurity took place in San Francisco as planned in late February, but big exhibitors such as IBM, Verizon, and AT&T Cybersecurity pulled out.

Conclusion

Stick back to the basics for now. The coronavirus is transmitted by the mist of the respiratory system, such as when someone sneezes or coughs around you. This also spreads influenza viruses and common cold viruses.

"The things you should do to keep yourself safe from coronavirus are stuff you should do on a regular basis," he says. T "The first thing you can do to prevent respiratory disease is to do good personal hygiene." The recommendation of the CDC to stop coronavirus (and other respiratory diseases):

· Wash your hands with soap or use an alcohol-containing hand sanitizer.

Sneeze and cough into your elbow tissue or crook. When you get mucus or saliva on your face, quickly wipe it out. Stop rubbing unwashed hands on your nose.

Stay away from people who are sick, especially those with respiratory and fever symptoms.

Stay home in case you're sick.

Clean the surfaces regularly and thoroughly with a disinfectant, such as countertops and door buttons.

Also, these are all basic rights that are meant to be standard everyday things. Extra security is not really required at this stage, including wearing medical masks, unless you have the virus or you are being tested for it. "As long as people don't sneeze, cough, or otherwise spill their respiratory excretions on you, you'd be fine," says Moorcroft.

Both the CDC and the U.S. Surgeon General have confirmed that medical face masks are not mandatory for people who are not at high risk (such as people who have already been in touch with people being treated for coronavirus), and that hand-washing is a stronger protection against the novel coronavirus.

Apart from the simple prevention of disease, a healthy immune system is the strongest (and only real) protection against disease. Your body is better capable of fending off diseases when the immune system is really humming, he says, so everyone should make an effort to form theirs to the top. Provide enough quality night sleep, stay hydrated, eliminate highly processed foods, and have enough micronutrients in your diet to do so.

CPSIA information can be obtained
at www.ICGtesting.com
Printed in the USA
BVHW080012011220
594480BV00008B/835